An Archaeology of Structural Violence

Cultural Heritage Studies

UNIVERSITY PRESS OF FLORIDA

Florida A&M University, Tallahassee
Florida Atlantic University, Boca Raton
Florida Gulf Coast University, Ft. Myers
Florida International University, Miami
Florida State University, Tallahassee
New College of Florida, Sarasota
University of Central Florida, Orlando
University of Florida, Gainesville
University of North Florida, Jacksonville
University of South Florida, Tampa
University of West Florida, Pensacola

AN ARCHAEOLOGY OF STRUCTURAL VIOLENCE

Life in a Twentieth-Century Coal Town

Michael P. Roller

Foreword by Paul A. Shackel

University Press of Florida
Gainesville · Tallahassee · Tampa · Boca Raton
Pensacola · Orlando · Miami · Jacksonville · Ft. Myers · Sarasota

Printed in the United States of America on acid-free paper

This book may be available in an electronic edition.

23 22 21 20 19 18 6 5 4 3 2 1

Library of Congress Cataloging-in-Publication Data
Names: Roller, Michael P., author. | Shackel, Paul A., author of foreword.
Title: An archaeology of structural violence : life in a twentieth-century coal town / Michael P. Roller ; foreword by Paul A. Shackel.
Other titles: Cultural heritage studies.
Description: Gainesville : University Press of Florida, 2018. | Series: Cultural heritage studies | Includes bibliographical references and index.
Identifiers: LCCN 2018014115 | ISBN 9780813056081 (cloth : alk. paper)
Subjects: LCSH: Coal miners—Violence against—Pennsylvania. | Anthracite coal industry—Pennsylvania. | Industrial archaeology—Pennsylvania.
Classification: LCC HD8039.M615 R65 2018 | DDC 303.608862233509748—dc23
LC record available at https://lccn.loc.gov_2018014115

The University Press of Florida is the scholarly publishing agency for the State University System of Florida, comprising Florida A&M University, Florida Atlantic University, Florida Gulf Coast University, Florida International University, Florida State University, New College of Florida, University of Central Florida, University of Florida, University of North Florida, University of South Florida, and University of West Florida.

University Press of Florida
15 Northwest 15th Street
Gainesville, FL 32611-2079
http://upress.ufl.edu

Dedicated to my parents, Peter and Akiko, whose amazing lives as immigrants and thoughtfulness as individuals have inspired my interests throughout

CONTENTS

List of Illustrations ix

Foreword xi

Acknowledgments xv

Introduction 1

1. An Archaeology of the Twentieth Century in the Anthracite Coal Region of Northeast Pennsylvania 7
2. History of the Anthracite Region, 1830 to 1897 31
3. The Memory of Violence: The Lattimer Massacre of 1897 51
4. Migration, Part 1: The Social and Material Landscape of the Company Town 77
5. The Archaeology of Machinic Mass Consumerism and the Logistics of the Factory Floor in Everyday Life 105
6. Migration, Part 2: Emancipation and Catastrophe 136
7. The Destructive Character: Renewal and Memory 163
8. Conclusion 189

Notes 207

References 209

Index 231

ILLUSTRATIONS

Map 1.1. The location of Hazleton and Lattimer 12
Map 4.1. Map from 1873 Beers Atlas with detail 82
Map 4.2. Company map from the 1920s 83
Map 4.3. Detail of 1885 map 88
Map 4.4. Detail from a company map dating to the second decade of the twentieth century 89

Figure 1.1. Company double houses along Main Street in Lattimer today 16
Figure 3.1. Bullets recovered from the site of the Lattimer Massacre 72
Figure 3.2. Map showing bullets identified in the survey of the Lattimer Massacre 73
Figure 3.3. Page from the *New York Evening Journal* 75
Figure 4.1. Architectural features 95
Figure 4.2. Structure floor plans 97
Figure 4.3. Remnants of the 1885 brick foundation and stone wall 98
Figure 5.1. Program from a flag-raising exercise held in Lattimer 107
Figure 5.2. Dr. West's Toothpaste cap and ad; Ben-Gay analgesic balm tube cap and ad 127
Figure 5.3. Artifacts representing new decorative technologies 128
Figure 6.1. Development of access roads to Pardeesville 146
Figure 6.2. A letter from the Community of Pardeesville Water Association 149

Figure 6.3. Procession of St. Nazarius 159
Figure 7.1. Real estate map of Church Street 166
Figure 7.2. Plans of the domiciles around Church Street 167
Figure 7.3. Graphic depiction of family movement 169
Figure 7.4. Image posted in the *Hazleton Standard Speaker* 171

Table 2.1. Immigrants to the anthracite region 44

FOREWORD

The Rise and Decay of Northern Appalachia

Michael Roller's *An Archaeology of Structural Violence: Life in a Twentieth-Century Coal Town* is a thought-provoking study that combines archaeology, historical research, and ethnography, and uses contemporary Marxist philosophy to develop a context to tell the story about work, labor, consumerism, and deindustrialization in northern Appalachia. While much has been written about central and southern Appalachia, focusing on the hardships of life and the social and political tensions that exist, many scholars have forgotten the northern reaches of this region and the long-term struggle of labor to co-exist with the exploitive nature of capitalism. Difficult histories are often missing from the national public memory; however, an archaeological investigation is an effective tool that can illuminate the tensions and conflicts that led to the destruction of an urban core in this Rust Belt region.

Roller's work is a unique archaeological response to the problems of dark or difficult heritage which charts the materiality of trauma and memory in the region. His study provides an overview of the legacy of the coal barons—the exploitation of workers and families and the degradation of the environment. In many ways, this difficult history has scarred the landscape and people, and today the region is facing many of the hardships that their ancestors faced several generations ago, such as poverty, underemployment, and general health and well-being deprivations. The dark and difficult history continues to plague the community and will continue to plague the region until people challenge the exploitive and destructive nature of capitalism.

Roller's work initially focuses on the little-known Lattimer Massacre, which occurred in 1897 in the anthracite region of northeastern Pennsylvania. Mine workers, mostly foreign born and unnaturalized, went on strike for better wages and better living conditions. According to Department of Labor statistics, the late–nineteenth century anthracite industry had one of the highest mortality rates. In fact, the Pennsylvania anthracite region led the world's coal mining regions in fatal accidents per 1,000 employees. Roller's archaeological survey of the massacre site and historical research of the event brings to light this relatively unknown conflict in American labor history. While a commemorative stone known as the "Rock of Remembrance" marks the general location of the massacre, the event is unusually absent from many historical texts on labor history, and it is also absent from the Commonwealth of Pennsylvania's school curriculum. However, Roller uses the event as a touchstone to explore the rise and decline of the anthracite coal industry and the development and eventual decline of the many coal communities in the region, tracing the impact of deindustrialization well into the twenty-first century.

Historical archaeologists have, for a long time, been grappling with various ways to comprehend and interpret the everyday material culture found at twentieth-century domestic sites. Roller advances the theme of violence into the mid- to late twentieth century and shows how softer forms of power and violence are held by capital and state over workers, citizens, and consumers. New definitions related to redevelopment, mass consumerism, governmentality, and biopolitics all play a role in this new form of power. Northern Appalachia continues to search for ways to revitalize infrastructure and community. Unions have lost clout and low paying wages now dominate the region's economy. New forms of racism have developed, dividing the local population.

Working with twentieth-century archaeological assemblages, Roller's study will become a key text for understanding the uses and meaning of modern material culture. The choice of theorists Roller uses to help explain these data is rather novel to the discipline. For instance, while propaganda experiments helped to develop a patriotic following during World War I, capitalists led by social scientists like Edward Bernays (nephew to Freud), successfully envisioned how new forms of propaganda could influence consumerism. Roller explains that businessman Edward Filene proposed in 1929 that mass production "holds possibilities of accomplishing for mankind all of the good that theoretical reformers or irrational radicals hope to secure by revolutionary means" (Filene 1929: 9, quoted in Ewen 1976: 87).

An Archaeology of Structural Violence: Life in a Twentieth-Century Coal Town provides an important framework for thinking about twentieth-century sites archaeology, focusing on issues like modernity, commodities, and labor under the broader context of capitalism and the mechanisms that make capitalism work so effectively. His tracing of the materiality of trauma in the region provides an avenue for understanding the region's difficult heritage and unraveling the roots of contemporary inequities. Roller's work has broader application when examining the destructive powers of capitalism. He shows how capitalism can easily abandon a community (or a region), leaving it underfunded and struggling to survive. Workers and communities feel abandoned as they endure injustices, and work toward justice through protest and resistance. It is a scenario that continues to be played out in the United States today.

Paul A. Shackel
Series Editor

ACKNOWLEDGMENTS

There are truly multitudes in this text who speak through their support, knowledge, and inspiration. I want to thank my colleagues and mentors at the University of Maryland who have shaped my ideas here through their own research, classes, and conversation. A special thanks goes to my colleague and coach, Paul Shackel, who always trusted me to "do the right thing" throughout the course of this project. I am grateful for the conversations I have had with Mark Leone which gave structure to this work. He initiated the pursuit of a critical archaeology in the 1980s and continues to push its boundaries. Thanks also goes to the reviewers of this manuscript, particularly Dan Sayers of American University, who helped me focus its sprawling wilderness into what you see today.

As archaeology is always only a team sport, thanks goes out to the many field school students, volunteers, and collaborators who helped in the gathering, documenting, and analyzing of this research. I could not have done this without you. There are too many to name, but a special thanks goes out to Justin Uehlein, Kristin Sullivan, Dan Sivilich and BRAVO, Camille Westmont, Myles Schaller, Katie Nyulassy, and Judy Joklik. Lastly, I will always be grateful to the community of collaborators and supporters of Hazleton, Lattimer, and Pardeesville, without whom this work would not have been possible. At minimum, they welcomed me as a guest to share, explore, and think through their history. A special thanks goes out to Joe and Barbara Michel, Angela Fierro, Maurice DeLorenzo and the DeLorenzo family, Robert Burczy, and Kent Jackson.

Finally, this book goes out to all those working in pursuit of justice and a better world. I hope what I have written here can inspire you.

Introduction

> The current amazement that the things we are experiencing are "still" possible in the twentieth century is not philosophical. This amazement is not the beginning of knowledge—unless it is the knowledge that the view of history that gives rise to it is untenable.
>
> *Walter Benjamin (1970: 259)*

This study is an attempt to conduct an archaeology of the twentieth century. As such, the material history and theoretical contexts recounted here are both locally specific and broadly generalizable. By shifting between micro- and macroscales of material life, I bring attention to the astonishing contradictions, ironies, and absences in our understandings of the century. Further, this archaeology of our recent past can show us how we got to where we are today politically, economically, and materially. Inherent to this approach is the critique that traditional ways of understanding this journey may be untenable.

In 2009, I joined my advisor Paul Shackel and fellow graduate student Kristin Sullivan on a research project on the outskirts of Hazleton, a city in the Anthracite Coal Region of Pennsylvania (Shackel et al. 2011). Our topic was a nearby town's dark heritage: the 1897 massacre of coal mine laborers from Southern and Eastern Europe. As we learned about the region, it became clear that the anti-immigrant sentiments that gave rise to the tragic event were not confined to the past. Today, residents largely descended from these immigrant populations in turn discriminate against the recent wave of immigrants from the Caribbean and Central America. Residents we interviewed sometimes evoked the past to challenge or validate this discrimination. Some justified their contempt by the positive example of their forebears. This ironic situation proved to be a fruitful mechanism to generate research questions. In time we came to ask, "In what ways do the racial tensions of the past echo those of the present? How do they differ?" and "How do the oppressed become the oppressors in time?"

In the course of my research, which expanded from an initial focus on the Lattimer Massacre to the objective violence of the region's company town and shanty landscapes, it became clear that merely comparing the past to the present would not produce a substantial understanding of either. Rather, any knowledge produced by this method would be an exercise in selective memory, the way local residents could use the past to justify or condemn their discrimination, but not to understand it. Moreover, I felt, such a history would be haunted by the traumatic ruptures and continuities of the past century.

Ultimately, the ethnographic research and the archaeological fieldwork provided a model for how to approach the region's past traumas. Conducting informal ethnography, we encountered the region's vernacular heritage, its modes of memory making, object collecting, and storytelling. One is struck by a hypertrophy of memory and forgetting when visiting the Anthracite Coal Region. The locals we met in homes, shops, and at our public archaeology digs shared with us stories, photographs, food, and objects. Along the way we encountered poetry, plays, and movie scripts recollecting the heritage of the region. Likewise, we were welcomed into homes, garages, and storefronts packed with vernacular displays of historic preservation and memory making.

The locals brought all of these to life with stories of hard times and a rich community life. They delivered their history to us with a nostalgia and melancholy that spoke to the great contradictions of the twentieth century, of great social advancement and material plenty, and of great violence and traumatic loss. In the context of these greater, albeit diffuse, traumas, the punctuating violence of events such as the Massacre seemed merely emblematic. In the anthracite region, "The tradition of all dead generations weighs like a nightmare on the brains of the living" (Marx 2003). Behind this collecting there was an unquenchable thirst for the recollection and affirmation of a lost "Golden Age." Through its memory work, the people of the region are continuously searching for the reasons behind the collapse of the local coal economy, the dissolution of community bonds, and relentless local government corruption. In time, I came to view these emphatic acts of collecting and organizing not as remnant traces of a fading way of life, but as contemporary, patterned collective responses to the trauma of the recent past.

Like much historical archaeology, the vernacular bricolage of memory work in the region concatenates the personal, the regional, and the political, overlapping individual, familial, and community memories into collective presentations of the past. Such a *glocalized* heritage seemed amenable to the methods of my discipline. With its primary evidence derived from *waste, ruins, surpluses,* and *redundancies*, archaeological tellings of history recognize these aspects of material life not simply for their contingency, but for their centrality within capitalist life across the passage of time. In this recognition, the great ironies of the twentieth century become patently visible, as an era marked by social and material progress, human achievement, and communal growth, but also by inequality, violence, social fragmentation, and nostalgia (Hobsbawm 1994).

To answer our research questions, I chose to use the evidence at hand to investigate the intervening years between the Massacre and today. Archaeology is used strategically, opportunistically, and dialectically: supported, amplified, and illuminated by archival and ethnographic research, spatial analysis, and social theory. Varied in its use of sources, it nonetheless returns again and again to the violence within everyday material life. Nearly every chapter explores the changing landscape of the company towns of Lattimer and Pardeesville and the shanties on their periphery, as well as the *leitmotif* of the Massacre.

THE CHAPTERS

Chapter 1 provides a brief historical and theoretical context from which to begin our study. It introduces the overall structure of the research with the concepts of subjective, systemic, and symbolic violence. A brief history of immigration and labor in the region leads to an exploration of the relationship between migrants and sovereign power. The next section expands this context by setting the stage for the present immigration issues. The last section discusses theoretical constructs and historical contexts explored in detail in the rest of the book. The often-antagonistic relationship between labor and immigration is presented in separate sections covering historic and contemporary milieu. Theoretical sections cover the relationship between sovereign boundaries and migrants in the past and present and the development, in the latter half of the twentieth century, of vertically integrated neoliberal and governmental subjectivities.

Chapter 2 provides a historical context of the anthracite region against the background of industrialization, modernity, and mass immigration. In this chapter the development of the anthracite industry is interminably linked to broader historical contingencies. Global depressions, political developments, and mass migrations are all demonstrated to have shaped the social and economic patterns in the region.

Chapter 3 describes the historical context of the Lattimer Massacre, followed by a recounting of the event and the trial that followed. Next there is an account of the archaeological survey conducted in the autumn of 2011 with the members of BRAVO (Battlefield Research and Volunteer Organization) of Monmouth, New Jersey. The archaeological account of the Lattimer Massacre described in this chapter exemplifies the capacity for archaeological evidence to be simultaneously concrete and ambiguous.

Chapter 4 examines the objective violence of the company town landscape, the setting for the antagonistic work and living conditions that gave rise to the Massacre. The chapter outlines the history of migration to the region, explicating the material and spatial factors governing "new immigrant" labor. Specifically, the chapter describes the development of shanty enclaves on the periphery of company-built housing during the turn of the twentieth century. This chapter concludes that these ethnic enclaves mark out new spaces of exception in the landscape of the town. Their presence materializes ownership's new dependence upon immigrant surplus labor pools and mechanized work processes to capitalize upon the fickle industry despite economic depressions, fluctuating markets, and an increasingly organized craft labor force.

Chapter 5 examines the motivations, characteristics, and materiality of what I term *machinic mass consumerism* on the national and local scale. The chapter outlines the rise of this phenomenon in the middle of the twentieth century, beginning with structural changes in political economy and national infrastructure following the First World War. Specifically, the chapter connects the efforts of industrialists and social scientists concerned with the effects of consumer democracy, including suppression of radical behavior and the profiteering of surplus production. The archaeological evidence used in the chapter comes from privy deposits and yard scatter from the yards of a shanty in Lattimer No. 2, excavated during the summer of 2013. These findings lead to a multiscalar analysis examining the implications of mass consumerism for the class positions of these most protypical of producers, immigrant laborers and their families.

Chapter 6 takes up the changing spatial landscape explored first in Chapter 4, in a much-altered political and economic environment. By the late 1930s, the anthracite coal industry was in decline. By 1938 the company divested itself of the company homes and the property around the shanty enclave. This is a period of emancipation for the town accompanied by the development of new forms of communal solidarity and identification. Attention to the material record suggests that this emancipation from industrial control signaled the development of new forms of collectively organized communal improvements. A countervailing influence also sets in with the development of contemporary ideologies of individuality, privacy, and ownership at odds with the historic materiality of the intimate communal spaces of the shanty enclave.

Chapter 7 examines several planning documents connected to redevelopment efforts of the last quarter of the twentieth century. Archaeology from this era also reveals drastic changes in the physical and social landscape of the town. I assert that urban renewal resulted not only in material and economic upheaval, but, more importantly, in a subjectivating process. Residents began to subject each other to bureaucratic demands suggesting that, at least tactically, they had adopted the language of renewal and management, enunciating a new community comprised of atomized individuals adopting entrepreneurial attitudes to space, labor, and governance. Material aspects of the landscape, which in previous eras served to draw together communal bonds, were recoded as nuisances and removed. At stake was the capacity for the materiality of landscape to remember, reproduce, and channel social relations in a manner responsive to the exigencies of uncertain economy.

Throughout the text, archaeological evidence is bound to the *longue durée* of twentieth-century history: the destructive cycles of capitalism entangled with modernist progress, empire, and class antagonisms. To do this I treat each stratum as implicitly *indexical* of what comes next: "a layered set of dangerous times" (Matthews 2002: 136). Each era reflects the same antagonisms of class and race, sovereign power and resistance, but inflected through different material conditions. Broadly, the intentions of this book are twofold. The first goal is to produce a particular history along the trajectory of a community in constant change. Second, my goal is to construct a nuanced periodicity of the forms of structural violence inside the everyday life of the twentieth century.

1

An Archaeology of the Twentieth Century in the Anthracite Coal Region of Northeast Pennsylvania

It includes the whole century. We took it as a sort of symbolic day on which is unleashed . . . all the conditioning, all the needed facts, just because the day itself has happened. But it does not only contain the past; it also contains the future.

Bernardo Bertolucci (1975: 12)

The Lattimer Massacre is among the bloodiest labor massacres of the nineteenth century, although it is almost entirely missing from most American history books today. In September of 1897 in the Anthracite Coal Region of Northeast Pennsylvania, a company-sponsored sheriff and a posse of local middle-class businessmen shot into a crowd of striking immigrant coal mine laborers. The posse killed nineteen instantly and injured more than forty more (Novak 1996; Pinkowski 1950). The marchers were among a diverse group of migrants from Southern and Eastern Europe striking for higher wages and better working and living conditions. Specifically, they sought the same treatment received by earlier generations of nativized migrant coal laborers from Northern and Western Europe. Carrying the American flag on their march, they sought these rights as aspiring citizens of their newly adopted country.

The Massacre, like the event described by Bernardo Bertolucci in his film *Novecento,* "does not only contain the past; it also contains the future" (1975: 12). I propose that the event and its contexts can serve as a kind of *critical constellation* of American social and political history, *indexing* contemporary and historical occurrences of class struggle, violence, profit capture, subjectivation, precarization, and racialization. While my research reflects the material history of the region spanning the late nineteenth until the end of the twentieth century, each chapter references the *leitmotif* of the Massacre of 1897 and the context of systemic and explicit violence that invests the event.

It is my goal to make this archaeological history both locally specific and broadly generalizable. To accomplish this task, I bind the analysis of the intervening years between the Massacre and the present to the destructive cycles of capitalism, modernist progress, Empire, and class antagonism that characterize the *novecento*, the twentieth century. Juxtaposing, grafting, and merging archaeological evidence with social theory, archival material, and ethnographic data yield an archaeological history greater than the sum of its parts. The account coalesces in an archaeological manner, stratigraphically. While the narrative ordering is chronological, the thematic subjects do not neatly obey the laws of superposition. As archaeological features are wont to do, the past and present sometimes intrude upon each other. Features, assemblages, and landscapes come entangled with heterogeneous strata of governmental and financial powers. Repeating motifs abound: ironies, surpluses, contradictions, ruptures. Like history itself, developments seen as progress in one era fulfill the conditions for ruination or putrefaction in another and vice versa.

To advance the theme of violence established by the Massacre, I illuminate the development in the mid- to the late twentieth century of softer forms of power and violence held by capital and state over workers, citizens, and consumers. Namely, I consider the political effects of systematized control afforded by the collusion of state and capital in the late twentieth century through factors such as infrastructure and redevelopment, mass consumerism, governmentality, biopolitics, and the boundaries of sovereign power. To do this demands my engagement with historical contexts and theoretical sources far beyond traditional Marxist accounts of class struggle, and the invocation of causal relationships beyond the dialectical. To accomplish this task, the text references the work of diverse theorists, including Walter Benjamin, Gilles Deleuze, Felix Guattari, Michel Foucault, Hannah Arendt, Slavoj Žižek, Giorgio Agamben, and Donna Haraway. This also implies taking advantage of the great strengths of archaeological evidence, which focuses upon surpluses, ruins, waste, and remainders as its subjects. The study dwells on surplus productions: peripheral settlements, surplus labor, states of exception, enjoyment, survival, endurance, and waste. Rather than a departure from an analysis of capitalist accumulation, a lingering on excess and surplus presents an alternative approach to understanding the realities of capitalism as they play out within the historical materialities of everyday life (Deleuze and Guattari 2009; Lazzarato 2014). These contradictory processes of capitalism are heightened by the introduction in the mid- to late-twentieth century of economic regimes of flexible accumulation and governmentality. To understand the nearly inscrutable penetration of financial logic into social life, the mobilizing of capital and labor markets, the introduction of neoliberal and governmental subjectivities, and the accelerated cycles of creation and destruction, demands critical structures that can make legible the manner in which they are embedded in everyday material life.

I leave it to the reader to decide if the result reads as comprehensive, complex, or simply messy. Ultimately, none is an excuse for neglecting the heterogeneous historical forces of the recent past that create our present. As I sit down to write this introduction in the summer of 2017, this history can reveal much about our present in a global political environment in which populist politics has normalized the discourse of nationalism, racism, anti-immigration, and coercive violence. Critical archaeology will not say anything of significance about our present until we can historicize the development of those recently enacted materialities, social structures, and deployments of power that, because of their proximity, resemble our present more than our past. It is neither too early, *nor too late*, to consider these within the domain of our critical analysis.

CONTEXTS FOR AN ARCHAEOLOGY OF MODERN VIOLENCE: PAST AND PRESENT

The Lattimer Massacre is a monumental event in its own right, although because of its obscurity, there is no single acknowledged narrative of its significance (Roller 2013). Historical and contemporary sources have interpreted the implications of the event in a variety of ways.[1] Revolutionary narratives consider the strike that precipitated the Massacre significant for the formation of class solidarity and collective bargaining rights in the region. Some of these accounts propose the event was the first step toward the development of class solidarity among miners and laborers of many ethnic backgrounds (Greene 1968). A massive increase in membership in the United Mine Workers union and the formation of local organization resulted. Newspapers worldwide cited the tragedy as an example of the oppressive apparatus forming between capital and the state during the Gilded Age. Other accounts lauded the posse for maintaining law and order.

Archaeologists can face difficulty interpreting and presenting the historical evidence of explicit episodes of violence between groups or individuals. Often, instances of violence occur rapidly and leave only ephemeral evidence. The circulation of these slight and often ambiguous materials can be distractingly provocative. When emotions or tensions are unresolved among descendent communities, these objects may attain a fetishistic significance or alternately, provide an extraordinary opportunity for engagement and healing. Furthermore, providing physical evidence of these dark episodes of history can spark reinvigorated attention. Such applications, however, may not provide a broader understanding of the *causes* of violence.

Without careful contextualization, evidence of discrete episodes of *subjective* violence can divert us from recognizing the liability of epochal *objective forms* that incite and justify violence (Žižek 2008: 1). By *subjective*, I mean the forms of explicit violence attributable to identifiable agents that populate the news: massacres and riots, abuse, murder, rape, and torture. These episodes occur between subjects, and have a history tied to the circumstances of particular times and places. Like the crime reported from four perspectives in Akira Kurosawa's film *Rashommon* (1950), they often occur within an overwhelming context of divergent or conflicting interests, perspectives, and justifications that can make it hard to tease out cause, culpability, and resolution.

Objective violence, divided into *systemic and symbolic* forms, are seemingly peaceful or stabilizing structures that serve as "the background which generates such outbursts" (Žižek 2008: 1; and see also González-Tennant 2018). Systemic forms of violence include the subjugation, manipulation, and control of material, social, political, or economic structures that compel or channel justice, attention, empathy, achievement, or resources away from those without power. Systemic violence is unconsciously entangled with symbolic forms of violence, manifested in the structures of language and meaning that subordinate groups to the interests of forces outside their control. Symbolic violence relies on the mischaracterization or simplification of the complex histories or cultures of marginalized individuals, communities, or groups to coerce them into determinist or essentialized economic or political destinies. Discriminatory systems such as racism, sexism, and gender discrimination all operate through the production of a constellation of internally consistent symbolic systems that amplify or distort aspects of material life to the disadvantage of a group, subjugating them into a relative position of "otherness" or marginalization to society at large.

Arguably, historical archaeologists are most comfortable and effective examining material history at the household or interpersonal scale. Archaeological analysis entails using the small and forgotten things to derive insight into broader truths that may otherwise be obscured or lost in the background. By rooting analysis in those situated objects, sites, and memories of subjective experience, archaeologists can make violence at all scales visible, at least as they arise in a particular historical constellation: a discrete time, place, and set of actors. By analyzing violent episodes through both objective and subjective systems, archeologists can draw out the causes of individual episodes of subjective violence and the objective forms they materialize (González-Tennant 2018). The goal is to reveal the underlying causes of violence that often hang in delicate balance in moments of danger (Matthews 2002).

The archaeological material used in this study originates from three excavations conducted between the autumn of 2011 and the summer of 2013. The first of these included an archaeological survey of the site of the 1897 Lattimer Massacre, in which a company-sponsored posse shot striking Eastern European mine laborers outside the town of Hazleton in the Anthracite Region of Northeast Pennsylvania (Map 1.1). I explore the broader systemic and symbolic context of this violence through an archaeological study of the region's company town landscapes, known colloquially as "patch towns." In these places, company owners structured a racialized social and labor hierarchy through the built environment. I propose that the objective violence of these landscapes facilitated the perpetration of acts of violence by labor and

Map 1.1. Map showing the location of Hazleton and Lattimer, where the research took place (map by author).

capital. This study documents the transformation of an enclave of shanties into a typical suburb over the course of a century. This analysis extends into periods of postindustry, and, finally, suburbanization, tracing the development of governmental and neoliberal subjectivities and the consequent dissolution of communal bonds. Using data typical of most archaeological studies, I explore this trajectory through changes in architecture, landscape, consumption, and spatial relationships.

The following sections provide theoretical and historical context for the investigation that follows. The first section discusses the materiality, context, and discourse of immigrant racialization in the late nineteenth and early twentieth centuries. It also includes specific information about the setting and early history of the archaeological sites examined. The section that follows continues the discussion of migrants in the context of the relationship between the sovereignty, borders, and labor throughout the twentieth century. In the next section, I discuss the context of immigration politics today. Lastly, I discuss the rise of neoliberalism and governmentality in the recent past. If these contexts form a kind of critical constellation to understand the following study, this last context is the navigational star to which this project directs us, an archaeological study of the development of soft forms of coercive power in the present.

Historic Context: Immigration and Industry in the Late Nineteenth Century

What struck me when I first surveyed the historic enclave of shanties in the town of Pardeesville, formerly Lattimer No. 2, was how much it looked like a typical suburban neighborhood. At first glance, it could be mistaken for any late–twentieth century suburb, with its lawns of closely cropped grass, backyard pools, and fenced-in yards. However, in the odd angles of the houses and their idiosyncratic footprints, I could still discern the organic intimacy of the enclave's historic landscape. On closer inspection, I recognized that many of the houses I had identified on company maps from the turn of the last century were still standing, now clad in aluminum siding. Nonetheless, a patina of deterioration reflected perhaps the most recent economic downturn. The landscape was a palimpsest, one that would only make sense if I could understand its development diachronically and as an organic whole. And seething somewhere below, I could sense the dark heritage of historical violence entrenched in this landscape.

Deeply entangled with the industrial history of the anthracite coalfields of Northeast Pennsylvania are the particularities of its geology and geography. Moreover, transnational flows of capital, commodities, and people have shaped it throughout its history. Geological deposits of anthracite coal, sandwiched between sheets of rock in strata of various thicknesses and declination, are an engineering challenge to excavate. Underground miners must chase coal seams across geological folding and faulting that defies the logic of aboveground space, sometimes diving precipitously below the water table. Mining operations required highly capitalized initial investment costs to mediate hazardous working conditions, including tremendous heat, flood waters, poisonous gases, explosions, and wall collapses. Tied to the vicissitudes of the global energy market, the value of coal was invariably unstable. To provide for the steep investment costs of this fickle industry and mediate its unpredictability, operators maintained large, highly stratified, and expendable forces of both highly skilled and unskilled labor.

Labor was drawn from the major waves of immigration entering the country, first from Western Europe in the early- to mid-nineteenth century, and then Eastern and Southern Europe throughout the late nineteenth and early twentieth centuries. The extent to which companies controlled the character and flow of in-migration to the region is controversial, although controlling interests undoubtedly benefitted from the destabilizing competition new immigrants inspired among the labor force (Barendse 1981: 7–8, 24–28; Brooks 1898). Racialized nationalities of each group came to occupy a place in a relatively positioned hierarchy. For a time, coal operators maintained this pool of surplus labor, which enabled them the logistical flexibility to respond to the pressures of unpredictable markets.

As in other industries, the new immigrants were at first assigned relatively less-skilled laboring roles and therefore earned considerably less (Dubofsky 1996; Gutman 1977). In turn, companies reified and exacerbated this racialized differentiation of new immigrants through the hierarchization of labor processes, the institutionalization and reinforcement of particular spatial and material conditions, and discriminatory litigation (Berthoff 1965; Mulrooney 1989; Novak 1996). In the earliest period after their arrival, between about 1875 and 1910, new immigrant workers occupied shanty enclaves on the margins of the ordered space of the company towns. This social marginalization was materialized in differentiated architecture and spatial patterning, consumer goods, and working and living conditions.

An increasingly rigid race- and class-based social hierarchy characterizes this period, as the relationship between white identity and the nation-state was in the process of legal and social codification (McKeown 2008; Ngai 1999, 2004a, 2004b). These developments affected not only the new migrants, but also long-marginalized groups including African Americans and Native Americans. Within this context, migrants from Eastern and Southern Europe, making up what was called the *New Immigration*, were coded for a time in American society as "inbetween peoples," occupying an epistemological grey area between whiteness and ethnic or racial otherness. Scholars of race and ethnicity hotly debate the trajectory of their eventual passage into whiteness. In question is the pace, sequence, and causality of this trajectory for individual groups (Barrett and Roediger 1997; Guglielmo 2003; Roediger 2005). The common agreement is that the passage was messy and contingent upon a variety of historic and political factors.

Capital and labor supported and exploited these forms of symbolic violence, which were popularly accepted and elaborated upon in nearly every aspect of social life. In a process described as "initial proletarianization," these unskilled workers fueled American industry, spreading across the country to work in factories, mills, sweatshops, mines, and other extractive industries and infrastructure development (Dubofsky 1991, 1996). Importantly, the region's coal economy integrated inequality and unemployment into its normal functioning, coupled with intergroup competition. Like many ventures of Gilded Age industrial production, this army of surplus labor mediated profit loss for owners by absorbing the impacts of natural or market variability (Harvey 2010; Walker 2008). For Marx, this is central to the optimized functioning of capitalism itself:

> The capitalist mode of production . . . forms a disposable industrial reserve army that belongs to capital quite as absolutely as if the latter had bred it at its own cost. Independently of the limits of the actual increase of population, it creates, for the changing needs of the self-expansion of capital, a mass of human material always ready for exploitation. (1977: 784)

Simply stated, a glutted labor market fueled wage competition. Poverty and frequent or persistent unemployment was a fact of living for

Figure 1.1. Company double houses along Main Street in Lattimer today showing alterations that give them a suburban-like appearance. Private owners have customized the houses, often purchasing both sides and combining them into a single-family home in conformity with late-modern suburban notions of space and privacy (photo by author).

the working poor of the region throughout its history. As a result, the realities of the Gilded Age for those at the bottom of the social hierarchy "wasn't so gilded" for laborers and their families in the anthracite region (Shackel and Roller 2012).

On a diachronic scale, tracing change in these spaces archaeologically reveals the material trajectory of migrants from excluded lives to American citizens across the twentieth century. Archaeologists frequently look to the built environment to find materializations of systemic and symbolic violence, demonstrating the power of landscapes to dictate ideological social behaviors through the imposition of differentiation, discipline, and homogenization (Epperson 2000; Leone 1995, 2005, 2010; Nassaney and Abel 2000; Sayers 2003; Shackel 1993). Architecture and landscape are especially effective as tools of domination as they are "relatively permanent, pervasive, and capable of fostering redundant messages" (Nassaney and Abel 2000: 254).

Still standing in the present, the majority of the towns of Lattimer Nos. 1 and 2 take the form of company-built double houses ordered in neat rows (Figure 1.1). Each house has a roughly identical footprint and is surrounded by a fenced-in yard of roughly comparable space. New immigrants were, for a time, confined to shantytowns at the periphery, their differentiation coded in distinctions of space, architecture, material goods, and infrastructure. On historic maps from the Lattimer Coal Company Records from between about 1878 and 1938, clusters of heterogeneously shaped and positioned structures develop at the periphery of the well-defined geometries of the company homes. In three distinct areas around Lattimer and Pardeesville (formerly Lattimer No. 2), myriad tiny structures appear, tightly clustered around narrow thoroughfares and alleyways. This configuration both echoes and contrasts with the broader spatial and aesthetic relationships between the company houses. Further archival research and ethnographic work revealed that these "irregular" areas were occupied by the latest waves of immigration to the region, from Southern and Eastern Europe. I theorize here that this differentiation and exclusion of populations such as these, reinforced through material means, explain how lethal violence was justified by the posse of middle-class businessmen in the event of the Lattimer Massacre.

Steps toward an Archaeology of the Twentieth Century, Part I: Labor, Migrants, and the Sovereign State

> Immigration has become . . . one of the most crucial sites where democratic states are put to the test. (Fassin 2011: 218)

The twentieth century is an era marked by great contradiction and irony. On the one hand, it is an era of unprecedented social and material progress, heightened communication, and cultural exchange. It is also defined by rising inequality, violence, social fragmentation, and the nostalgia and melancholy of excessive trauma and loss (Hobsbawm 1994). Particularly during times of economic recession and stress, the most vulnerable suffer most deeply. This includes groups marginalized by race, ethnicity, gender, and sex. It also includes recent immigrants, the working class, the elderly, and the very young, who all may suffer under the burden of precarity, vulnerability, and debt. During times of strife, internal and transnational, those in power have historically targeted immigrants, using them as scapegoats to explain away the failures of national economic or political policies.

That migrants are invariably blamed for the nation's problems cannot be explained as a simple matter of convenience. That politicians and citizens provoke debate over immigration reform whenever a nation encounters undue amounts of internal or external stress is unsurprising. Migrants have always held a formative relationship to the development of sovereign power, national identity, and the structure of national subjects. Moreover, in their vulnerability to law and extrajudicial practice, the migrant exemplifies the precarious condition of all subjects of the state by illuminating the ideological gaps between political identity and bare human life (Agamben 1998; Kristeva 1991).

State power over its subjects is classified as sovereignty: the authority to declare life or death, to suspend the rule of law, or to act outside of it. Sovereign power is intrinsically spatial, suggests Giorgio Agamben in *Homo Sacer* (1998), through its capacity to define the boundaries, metaphorical and physical, between those included and excluded from the social benefits and protections of the state. As a self-justifying act that tactically produces itself, the lawmaking act of sovereignty is always already "necessarily and intimately bound up with" violence (Benjamin 1978: 295). The subject defined through sovereignty is the figure of *homo sacer*, or *bare life*, upon which power acts (Agamben 1998). The figure of bare life embodies a kind of raw human identity, an exceptional characterlessness occupying the contradictory position of defining the nation's boundaries by its exclusion. In classical models of sovereignty, bare life serves as a kind of foil for the monarch, an exceptional force outside of law, but by whom the law is given shape and direction.

The regulation and facilitation of the movement of bodies across borders are inseparable components of the same process. "Contact," asserts historian Adam McKeown, "generates not only assimilation and convergence, but also new ways for people to distinguish themselves from each other and from what they perceive to be a homogenizing universalism" (2008: 5). In late modernity, bare life came to be defined by those who fell between the cracks of a world defined entirely by the interlocking territory of state sovereignty. During and after the Second World War, the atrocities and protections perpetrated by totalitarian regimes were the culmination of the intermingled growth of

Late Modernity, industrial capitalism, and the nation-state. The flood of war refugees, adrift and shorn of the protections of national belonging, represented an existential threat to the notion of the nation-state as the sole ordering force (Agamben 1998; Arendt 1958). Nations were, "for the first time confronted with people who had indeed lost all other qualities and specific relationships—except that they were still human" (Arendt 1958: 299). Through their encounter with the flood of wartime refugees across borders, an international discourse regarding universal rights of humankind was born, inspired by the search for a quality in the human condition that exists outside of the definitions furnished by state sovereignty (Agamben 1998; 126–135; Arendt 1994 [1943]).

If migrants already signal an exceptional biopolitical subjectivity on the margins of the nation-state, the proletariat represents another form of disarticulated precarious agent in history, cast adrift with the dissolution of classical forms of power. However, their historical condition also reflects the contrary position of serving economy by supplying necessary labor. Marx anticipated the condition of alienated subjectivities in his analysis of primitive accumulation and the "double freedom" of the proletariat (de Boever 2009; Marx 1977: 874). In his early essays, Marx describes the proletarianization of peasants emancipated from feudal systems as making up a class of *vogelfrei*, which can be literally translated as, "free as a bird." This translation resonates with a notion of the conclusion of feudal arrangements as an emancipation.

A reckoning of such an interpretation with the materialist historical narratives of Marx's work suggests a rather more complicated and compelling understanding of the concept of *vogelfrei*. Moreover, it is one that holds a close affinity to the concept of bare life outlined by Agamben (de Boever 2009). Marx documented the trajectory of European peasantry out of feudal systems as going from bad to worse: emancipated families were "freed" to sell their bodily energies into industrial labor. In fact, the etymology of the term *vogelfrei* has a double meaning, quite divergent from any implications of emancipation. In later works its usage is better translated as "the outlawed," "homeless," or better yet, "free for the birds to be eaten, since the body of an outlawed person could not be buried" (de Boever 2009: 264). The emancipated were freed, yes, unburdened, yes, but ironically also left to vulnerability and precarity, always hoping for the promised "forty acres and a mule" (de Boever 2009). In their condition of precarity and homelessness, there is a contiguity in the bare life of workers and migrants who are freed of certain constraints only to be cast adrift in a world in which there are few cracks between sovereign borders or work arrangements.

And what of the migrant laborer, whose conditions of freedom and marginality are doubled? In the context of state economies built upon industrial capitalism, the conflicting patterns of national identity and the insistent demand for a supply of unskilled labor place migrants in the contradictory position of being both appropriated and restricted. History has shown that, as a result, inconsistency or ambiguity has defined their treatment, particularly for national economies that depend upon migrant labor for profit (Camp 2011; Coutin 1995; Fassin 2011; McKeown 2008; Ngai 2004). In the twentieth century in the United States, the group with perhaps the longest and most tumultuous relationship with these conflicting strains of dependence and rejection are Mexicans and Central Americans, who have seen the border between their country and the United States open and close with startling regularity (Chavez 2008; Longazel and Fleury-Steiner 2011; Ngai 2004).

The sovereign power to draw boundaries between excluded and included comes into play in both the subjective violence of the Massacre, and in the objective violence of the divided landscapes and split work regimes of the anthracite coal industry.

Immigration History Today: The Irony of the Past and Present in Juxtaposition

Grand historical narratives are intractably streamlined affairs, composed as much of selective forgetting as remembering. Linear narratives such as evolutionary models of ideological progress or vulgar economic determinism may illuminate some aspects of broad history, while obscuring others. Outside of narrative, history inevitably unfolds in contradictory and ironic ways. Donna Haraway (1990: 149) suggests that for her, only through the formation of an "ironic political myth" could she find a way to account for the contradictory elements of history. Irony as historical method is, "about contradictions that do not resolve into larger wholes, even dialectically, about the tensions of holding incompatible things together because both or all are necessary and true." Fundamentally, it is about tracing the complexities of history that grand narratives obscure.

For Haraway, "Irony is about humor and serious play. It is also a rhetorical strategy and a political method" (1990: 149). Juxtaposing the era of the Massacre with the ethnographic present reveals ironies that challenge a linear narrative of social progress. Immigration politics in Hazleton today echo the violent episodes of its deep history. As in the past, these conflicts are deeply embedded in national immigration debates and a globalized political economy. In Hazleton in the 1990s, the municipality struggled to advance its economic position in the context of a neoliberal order of unregulated market competition, confronting forces outside the control or understanding of its citizens or municipal authorities (Longazel 2016: 17–19; Varsanyi 2011).

Hazleton's economy struggled to stay afloat in the years after the anthracite coal industry essentially ended in the 1950s. As in other small towns across the country, it suffered the persistent traumas of late capitalism that deeply affect the economic and social life of rural, postindustrial, or otherwise peripheral communities: deindustrialization, disinvestment, stretched municipal budgets, youth outmigration, persistent unemployment, and aging populations (Bodnar 1992; Deasy and Griess 1965; Dublin and Licht 2005; Theodori and Theodori 2014). For a time, the CAN DO Corporation (Community Area New Development Organization), a community effort to court industry, successfully brought light manufacturing jobs to the region. But this success began to wane as the national economic climate restructured in the 1970s, turning toward free trade, public-sector austerity, low inflation, and flexible labor markets (Harvey 1989, 2005; Varsanyi 2011). Reflecting the intensified mobility of global industry, companies that could move to labor markets that were cheaper than Hazleton promptly decamped, often for the American South or overseas. Between the 1970s and the 2000s, Hazleton lost as many as 20,000 jobs, mainly in the manufacturing sector (Longazel 2014: 585). Like many small industrial towns throughout the country, Hazleton was powerless to negotiate the retention of stable, high-paying jobs in manufacturing, much less the institutional support of worker organization (Bluestone and Harrison 1982; Cowie 2001). Local labor regimes bore the costs of both global competition and local acquiescence in exchange for ephemeral employment arrangements (Fleury-Steiner and Longazel 2010; Longazel 2016).

In the 1990s, the state of Pennsylvania began a program to create large tax-free zones to attract new industry to its blighted regions. Hazleton took advantage of this program, successfully competing with other municipalities to acquire the largest acreage of these incentives (Longazel 2016: 19). These developments, in turn, brought a variety of warehousing industries to the city, which sits conveniently at the intersection of major highway routes I-81 and I-80. Because of competition, the town could only acquire temporary jobs, which paid lower wages and exposed workers to dangerous working conditions. Many firms reportedly also had a reputation for exploiting immigrant labor (Longazel 2016: 19).

Central American and Caribbean immigrants began migrating to Hazleton, attracted by the work and low cost of living. Many had been priced out of gentrifying East Coast urban neighborhoods. Downtown Hazleton, largely abandoned, offered new immigrants willing to endure the lack of services a substantially cheaper cost of living. Storefronts were often empty and could be rented cheaply. Employers eagerly sought young or energetic employees willing to endure the economic and physical conditions they offered. The Hispanic population in Luzerne County, where Hazleton is located, grew from 3,713 to 23,125 between 2000 and 2011, a 523 percent increase (Pew Research Center 2013).

By the early 2000s tensions began to rise between the local residents and the newest immigrants, associated with the growing national concerns about border security which arose in the wake of 9/11. In 2006 Hazleton was the first municipality to pass laws aimed at limiting immigration. Known as the Illegal Immigration Relief Act (IIRA) and the Official English Ordinance, these ordinances pursued the localization of immigration policy, claiming that national law was not effective. Hazleton, followed by many other localities, pursued the criminalization not only of undocumented migrants, but also those who employed or housed them (Dick and Wirtz 2011; Kobialka 2012; McKanders 2007). Although the IIRA in Hazleton was never enforced, and was ruled unconstitutional by the Supreme Court in 2013, it immediately altered the character of interactions between long-term residents and recent immigrants. Through discursive practices of politic demagoguery and media, recent immigrants were racialized as illegal, unassimilable, lazy, criminal, and stubbornly refusing to learn English (Dick and Wirtz 2011; Kobialka 2012; Longazel 2013a).

An informant, a naturalized citizen originally from South America who has lived in Hazleton since the late 1980s, described these discursive and justificatory effects of the laws:

> From 1989 to 2005 we didn't have any problem with immigration. Absolutely nothing. 2005 when the Hispanics started moving a lot to this town is when it started, by political interests trying to stop the immigration of people using [terms like] "illegal," "terror," and the city took the occasion to approve an ordinance called the IIRA . . . when they approved that ordinance the people that were probably under the bed or in the closet came out: they felt free to say "I don't want you here. You speak English with an accent or you don't speak English at all or you don't have papers." . . . The ordinance gave the people the power to be racist, to be less compassionate. And it divided the city. The city before that was just like any other city. No problem. But after that ordinance the city started being divided. People that were Latino and that were foreign were hated. (AA, interview, July 2011)

The laws, though never implemented, were nonetheless effective in other ways. The laws normalized the symbolic violence concealed by language, distilling, channeling and unleashing the tensions of the residents of the town.

The recent immigrants to Hazleton face discrimination similar to what Eastern and Southern Europeans received a century before. At times, too, Hazleton residents ironically evoke the Massacre or their immigrant heritage to reflect upon the sacrifices of earlier generations. Locals evoke history to convey both positive and negative comparisons to the efforts of recent immigrants to integrate into the region's economic and social life. In an example from a Hazleton-area newspaper editorial from 2009, a commentator writes:

> Read what happened to your ancestors in the Lattimer Massacre. Read about the reaction of native Hazletonians in this newspaper

> to that horrendous mass murder. . . . Accept all the people of Hazleton as one people. (*Hazleton Standard Speaker*, October 7, 2009)

At other times, the reverse comparison is made. In an interview recorded by our project in 2009, an ethnographic informant compared past and present immigration:

> There were a lot of immigrants coming then; there're a lot of immigrants coming now. But I think . . . the trouble is a good number . . . of the immigrants that came off of Ellis Island . . . were coming because they were hard workers, [and] wherever they were coming from wasn't the best, and they knew that America was the promised land, and they could make a new start. Whereas today I think a lot of people come here because they know how easy it is to live off of the government. . . . They're not . . . trying to get their papers like . . . Italians or Germans or Irish did. They're coming here because . . . someone said "Hey, you could live in Hazleton for very very cheap prices and you don't even have to tell them that you're there." (BM, interview, March 2010)

Residents such as this justify their distinction between the hard-working immigrants of the past and those of today.

Scholars have pointed out that, ironically, the same neoliberal economic restructuring that creates ultracompetitive labor markets dependent upon immigrant labor also produces the economic conditions that justify their restriction by locals (Calavita 1994; Longazel and Fleury-Steiner 2011). It was in fact the onset of post-Fordist economics in the mid-1970s, which decimated the economies of many American towns built on industry. Many industries decamped production from the expensive labor markets of the Northeast and Midwest almost overnight because of radical deregulation and the facilitating effects of flexible and mobile production technologies and mass transportation networks. The competition that resulted drove employers to reduce production costs by mechanizing and de-skilling work, cutting wages, benefits, stability, and work conditions. Only immigrants were willing to take on many of these precarious work conditions.

Neoliberalism is also a *subjectivating* enterprise in that it fosters an environment in which particular kinds of individuals can thrive, while strategies for advancement suited to past conditions are no longer effective. The penetration of economic logic to individual lives is concretely observable in the ways that supporters of immigration restriction use neoliberal values to justify discriminatory policy. In place of the more explicit racialized language of nineteenth-century immigration restriction, today groups are targeted for their failure to conform to neoliberal values of entrepreneurship, self-sufficiency, compliance, individual striving, and freedom from collective enterprise.

Steps toward an Archaeology of the Twentieth Century, Part II: Subjectivation, Neoliberalism, and Governmentality

> Before the law sits a gatekeeper. To this gatekeeper comes a man from the country who asks to gain entry into the law. . . . "It is possible," says the gatekeeper, "but not now." . . . The man from the country has not expected such difficulties: the law should always be accessible for everyone, he thinks. (Franz Kafka 1976: 3)

In a parable by Franz Kafka, a man from the country petitions access to the law, which, in his words, "should always be accessible for everyone." He awaits admittance for an eternity, even providing bribes to the gatekeeper, to no avail. Just before he dies, he inquires as to why there are no others waiting with him. The gatekeeper, who has blocked his entry all these years, reveals that the gate was built only for the man from the country, and that when he passes away it will be closed off. The gatekeeper tells the man that he has taken these bribes all along only to reassure him that he has done all he could have done to enter.

My interpretation of Kafka's parable is that it reflects how the desires of nineteenth-century subjects have been exactingly reflected back into the political economy and soft forms of political control we experience today. The individualized fate of the man from the country, before the gate constructed specifically to block him, is an ironic perversion of his belief that the law be available to everyone and at all times. Through his stories featuring the confrontation between individuals and modern bureaucracies, Kafka was the first to "articulate [the] paradoxical status of the subject before subjectivation" (Žižek 1989: 205). His tales reflect observations of the powerful new forms of objective violence he perceived within the labyrinthine procedures of modern bureaucracies.

The central irony of the parable propels the research interpretations here: if we are not careful, even a misplaced desire for justice can lead to a greater oppression.

The fate of the man from the country presages the particular forms of *unfreedom* exhibited by neoliberal individuals, free from all social constraints but for the most binding: that of an individualized debt uncontestable by collective rejoinder. He finds himself standing alone before the law, and although the gate is built especially for him, it is stripped of the individual qualities that might attain to a personal or group history. Like a passport or database entry, it is individualized only to the reductive extent that its data can be collected and traced.

Historian Adam McKeown (2008) notes that the capacity of the state to generate individualized political subjects arises out of the creation of border restrictions developed at the turn of the last century. This capacity is mediated through the materiality of individualized passports, visas, and border checks. The development came not only from a governmental effort to identify favored and unfavored citizens but also from an attempt to separate the individual from the collectivity that presented the threat of collectivized combination among the nation's immigrant labor force. The essential unit of movement in this equation was the individual, separated from collectivities of kinship, informal social networks, labor contracting, or other institutions that frequently accompanied past immigration (McKeown 2008: 11). The requirements to render individuals distinguishable resulted in the development of techniques of classification in two different forms (McKeown 2008: 12):

> the characterization of each person as a unique, physical object, and the development of technical and testable categories of status, occupation, and family that could be standardized within and across nations. Technologies like photography, fingerprinting, and anthropometric measurement helped to define persons as unique objects.

The neutral language of these bureaucratic procedures, moreover, replaced overtly racist identifications, just as it served to place the burden of disciplined nationalized behavior upon individual subjects rather than collectivities.

If the border control policies of the early twentieth century laid the ideological groundwork for the individuation of subjects, the rise of neoliberalism and governmentality in the latter end of the century is its culmination. Neoliberalism is not so much a transformation in ideology as a transformation of the very "effects and conditions" by which ideology is materialized, practiced, and concentrated (Read 2009: 26). Beyond mere discourse or falsification of reality, it is a form of normative reason that holds the financial market as the source of rationalizing logic framing, "every sphere and activity, from mothering to mating, from learning to criminality, from planning one's family to planning one's death" (Brown 2015: 67). Thereby, neoliberal individuals identify themselves through market logic, "encouraged to cultivate themselves as autonomous, self-interested individuals, and to view their resources and aptitudes as human capital for investment and return" (Binkley 2009: 62). Within this context, the neoliberal subject is, on the one hand, not coerced or disciplined, but, on the other hand, "obliged to be free," while also pressed into a constant state of political exception. Understood in opposition to social or co-constituted relations of kinship, community, ethnicity, and so on, these formations produce the contradictory position of individuals, "manufactured . . . from within and against their original communities" (Matthews 2010: 11). Moreover, these "distinct individuals" come to believe they freely choose the nature of the social relations they enter.

Governmentality is the political counterpart to neoliberalism, which also traces its form from an acute vertical integration between the state and the individual. As such, it is the counterpart to sovereign power (discussed above). As opposed to explicit displays of sovereign power, however, governmentality, "addresses the historical specificities of government, its style, its premises, and its instrumentation" (Palus 2010: 5). The goal of governmentality is that of security, not domination or coercion. This, however, does not preclude an exercise of power. Through the adoption of mentalities of rational organization and welfare, populations securely fashion their own lives in a way that nonetheless consolidates the power of government and capital (de Boever 2009; Foucault 2009).

To take this discussion full circle, Wendy Brown (2015: 108) proposes that a "ghostly repetition of the ironic 'double freedom'" of Marx's proletariat arises in the "double freedom" afforded by contemporary neoliberalism. Neoliberalism is founded on principals of freedom: markets, individuals, and politicians are all emancipated from constraints to operate as entrepreneurs in open competition. The neoliberal individual has been freed from social and political constraints, but also from the protections and rights afforded by classical forms of sovereignty or solidarity. On this broader scale, the programmatic aspects of neoliberalism are a structural effort to "put into question all collective arrangements capable of obstructing the logic of the pure market." This includes all collectivities that might mediate the relationship between governmental or capitalist logic and free individuals (Bourdieu 1998: 96). Tethered vertically to market logic along with the priorities of governance, Brown contends, individuals lose the capacity to make demands or organize their lives based upon any other logic than that of the market (2015). Neoliberal subjects need not be governed but are constrained to act and think within the mentality of government.

Arguably, such forms of disciplinary soft power produce a kind of symbolic and systemic violence far more violent than the coercive capital-labor tensions of previous generations. These dispersed and largely indiscernible structures of society not only provide the grounds to justify physical violence, but coerce and constrain individuals into forms of subjectivity, relation, and alienation (Sayers 2003, 2014; Žižek 2008). In essence, they yield alienated communities of autonomous individuals tied to regional governance rather than communal structures of support, communication, and meaning. Viewed through these tactical modes of power, we can conceive of the developments of the twentieth century beyond the obfuscating narratives of grand ideological clashes, binary oppositions, and charismatic actors that traditionally define readings of the century. Agamben (1998: 121) advises that reinterpreting the history of the twentieth century through the power effects of biopolitics leads to a recognition of the "contiguity between mass democracy and totalitarian states":

> It is almost as if, starting from a certain point, every decisive political event were double sided: the spaces, the liberties, and the rights won by individuals in their conflicts with central powers always simultaneously prepared a tacit but increasing inscription of individuals' lives within the state order, thus offering a new

> and more dreadful foundation for the very sovereign power from which they wanted liberation.

Such a theoretical position makes legible the contradictory, and often ironic, developments of the century. Through this reconceptualization we can focus on the ontological effects of modern materiality, depoliticized rational administration, and the penetration of financial logic into everyday life. Cleared of the obfuscation of historicization, we now have a foundation from which to begin an archaeology.

THE DESTRUCTIVE CHARACTER

> The destructive character knows only one watchword: make room. And only one activity: clearing away. His need for fresh air and open space is stronger than any hatred.
>
> . . .
>
> The destructive character is young and cheerful. For destroying rejuvenates, because it clears away the traces of our own age; it cheers, because everything cleared away means to the destroyer a complete reduction, indeed a rooting out, out of his own condition. (Benjamin 1978 [1931]: 301–303)

This fragment from Walter Benjamin comes from a text entitled "The Destructive Character." In this text, the paradoxical character of destruction is illuminated. Benjamin likens destruction to the creativity of progress, which makes room for the future only by obliterating the past. It is an illuminating allegory for the ironic modes of destruction wrought by modernity and capitalism in the twentieth century.

This study reflects the growing scholarship on *difficult, dissonant,* or *dark heritage* (Logan and Reeve 2009; McAtackney 2013; MacDonald 2010; Tunbridge and Ashworth 1996). But it also responds creatively to the dissonant heritage it evokes by offering a critical archaeological response. By confronting our inheritance of the materials and memories of difficult aspects of our past, it strives to resolve antagonisms and contradictions in the present. Paul Shackel (2013: 318) notes that exploring difficult heritage, such as historic episodes of subjective and objective violence, is an important social exercise of national memory making. "By bringing to light the conditions of the past and connecting these issues to the present, we can make some of these difficult histories a platform from which to discuss the continued prevalence of these inequities" (2013: 318). Confronting difficult histories can "illuminate the roots of contemporary social, economic and political injustices" (2013: 317). The often-neglected heritage of immigrant labor struggle for equality in the workplace and the home is an area of critical importance to expanding civic discourse. The absence of these struggles from civic memory means public discussions of the struggle for worker protections and wealth inequality lack the vocabulary or grammar of historic examples. Memorializing working-class history, both tragic and victorious, is one way to ensure that critical debate surrounding these issues continues (Little and Shackel 2014; Shackel 2009; Smith et al. 2011; Shackel 2013).

Archaeology is a journey into the unknown. Its primary material evidence issues forth from between the cracks of historical narrative. As such, I propose that the flattened ontology of a history based on materials can resolve an understanding of a dissonant history. Critical methods demand that every examination of the past proceed self-reflexively (Leone et al. 1987). In the interpretation I present here, this entails a recognition of post- and predepositional temporalities, materials, and activities that may affect any given interpretation.

Echoing Palus (2010), I suggest that, given the drastic transformations of twentieth-century life, the archaeological models, theories, or contexts used to understand the behavior of the nineteenth century and earlier are not appropriate to understand materials from the recent past. By extension, we will fail to understand our contemporary moment or present alternatives for our future until we can approach the recent past archaeologically. To accomplish this requires developing a nuanced materialist periodicity of the twentieth century (Gonzalez-Ruibal 2008; Rathje et al. 2001). This is perhaps the most important endeavor for a critical archaeology today.

2

History of the Anthracite Region, 1830 to 1897

If we sum up briefly the results of the work in the mines, we find . . . that through prolonged childhood on the one hand and premature age on the other, that period of life in which the human being is in full possession of his powers, the period of manhood, is greatly shortened while the life in general is below average. This, too, on the debit side of the bourgeoisie's reckoning!

Friedrich Engels (1969: 273)

For his 1844 report on the working class of England, Friedrich Engels traveled to Northern England to survey conditions in the coal mining industry. He reported on execrable working and living conditions including the *cottage* and *trucking* systems, which bound miners and their families in debt to company housing and credit. It was at this time that he wrote the powerful summation written above. It is likely some of the same workers surveyed by Engels later moved to the anthracite region of Pennsylvania in the period shortly after his visit, unfortunately meeting similar working and living conditions. The class structures of the industrial era were, after all, transnational, and crossing borders provided only a change of scenery and perhaps the diversity of new neighbors.

As twisted and capricious as its coal seams, the anthracite industry of northeast Pennsylvania compresses great complexity into a relatively short history. Marked by the constant movement of people, capital, politics, and materials, the region's remote setting served as both conduit and obstacle for its turbulent relationship to metropolitan, national, and global developments. Throughout the period of its early development, themes of industrialization, transnational immigration, and worker proletarianization all contributed to social life in the region.

INDUSTRIALIZATION AND ANTHRACITE

The technological achievements of early industrial capitalism seemed to produce a limitless mastery over matter, time, and space. The pace of manufacturing accelerated with no apparent end. Consistent and efficient sources of energy such as steam and electricity, dependent upon the burning of coal, replaced or augmented wood, water, and human labor. This led to the concentration of manufacturing in large-scale facilities such as factory complexes. These material transformations seem, on one level, to manufacture a world of efficiency, predictability, and plenty. On closer examination, this surface was phantasmagorical.

By the late nineteenth century authors Mark Twain and Charles Dudley (1873) would dub their era "the Gilded Age," implying that the surface-thin veneer of advancement concealed a baser metal. Twain astutely perceived the toll industrial mechanization wrought on the nation's workers in the form of inequality, systemic violence, and the undermining of democracy, as well as the destruction of the natural environment.

For the coal industry the material transformations of industrialization and modernity stand in frank juxtaposition to the violent and dehumanizing social effects of the industry. Coal is the "mucky material bottom" underwriting the energy, light, and movement of modernity (Arnold 2014: 266–267).

Anthracite coal provided the key factor jumpstarting American industrial expansion. Until the late 1830s, the United States lagged far behind British developments, partially for lack of efficient, centralized sources of energy. Providing a dependable, inexpensive, powerful source of energy relatively close to the metropolitan northeast, domestic coal production weaned American dependence on expensive foreign imports. Iron and steel production soared, no longer limited to isolated areas where ore and trees coincided in plentiful supply (Chandler 1972: 146–147, 159). The production of high-quality domestic steel, in turn, accelerated the production of durable and efficient machinery (Chandler 1972: 146, 165–174) and new steel-based industrial landscape, material culture, and architectural forms. The elements of steel and steam, underwritten by coal, came together in perhaps the most important development of the period, the railroad network, which in turn served to transport raw materials such as coal, ore, and wood into industrial centers, and finished products to a growing national market (Arnold 2014; Cronan 1991; Schivelbusch 1977).

In fact, the turbulent development of the coal industry, and its dirty, inefficient and potentially lethal labor processes of extraction, reveal the flip side of Gilded Age optimism (Arnold 2014; Johnson 2010). The rise of mass manufacturing characterized by urbanized, mechanized, segmented labor processes arguably stems from the specific qualities of coal as an energy source. In the case of the proletarianization and mechanization of labor, capitalists and managers took advantage of the predictability and intensity of the energy produced from coal to compress production schedules and de-skill work. Arnold (2014) argues that, "it was coal that enabled managers to strip craftsmen of their status; coal freed intellectuals and corporate managers to embrace the seemingly more certain science of abstract economic theory over the seemingly less certain alchemy of democratic politics . . . coal helped shift the nation . . . to a faster paced, more dynamic, modern society" (2014: 2). In essence, the segmenting of work regimens into de-skilled processes reinforced the divisions between wage laborers, managers, and monopolistic wealth (Chandler 1972: 177–178).

Formation of the Anthracite Industry (1820–1870)

More than metaphorical, the twisted materiality of the region's geology contributes an entangling complexity to capitalist social industrial relations. The presence of anthracite coal in the northeast corner of Pennsylvania was known as early as the late eighteenth century, but difficult geological and geographical conditions prohibited its efficient extraction and movement until an economic, social, and material infrastructure for its production, consumption, and transportation was established in the middle of the nineteenth century (Dublin and Licht 2005: 10–34). Anthracite coal, composed of nearly pure carbon, requires a high temperature to burn, but when lit produces a high-temperature flame with very little smoke owing to its relative lack of impurities. Despite this superior quality, establishing and maintaining a mass market for anthracite would prove difficult throughout its history. Domestic and industrial consumption of anthracite would not commence until industry interests developed specialized furnaces and hearths to burn it, and convinced the public of the utility and thrift of its adoption (Dublin and Licht 2005: 10; Itter 1934: 28–29). The development of a reengineered grate for burning anthracite in fireplaces is attributed to Wilkes-Barre industrialist Judge Jesse Fell in 1808, paving the way for the exploitation of a massive domestic-use market (Dublin and Licht 2005: 11).

Anthracite coal occupies four distinct fields running roughly southwest to northeast across six counties and about 480 acres of northeast Pennsylvania. Comprised of the Northern, Eastern Middle, Western Middle, and Southern fields, each region developed with a particular set of social and economic arrangements, extractive techniques, labor relationships, and integrated capital interests (Dublin and Licht 2005: 10). Each region transported its coal to different markets based upon geographical utility, transportation logistics, and entrepreneurial relationships. Southern fields shipped to Reading, Harrisburg, and Philadelphia, while the Middle fields shipped to New Jersey, New York, and by steamship to Boston. Throughout the region's history, this heterogeneity of practices, capital interests, and labor demographics would restrict unified actions on the part of both labor and capital to standardize procedures, organizations, or agreements (Aurand 2003: 14–22; Dublin and Licht 2005: 10).

The geographical and geological provenance of anthracite complicated the laborious processes of its extraction and transportation. Unlike English and Virginian coal, which are often located on the shores of rivers or oceans, anthracite is found within and between isolated mountain ranges in diagonally stratified beds. In the early days of industry, massive capitalization was required to construct the network of canals and railroads to connect mineral sources to markets in the industrial centers to the north and east. By the 1830s, investors had completed three major canals in the region, expanded eventually to a total of six. Prior to 1850 the anthracite canals ranked among the "most heavily capitalized privately owned corporations in the nation" (Powell 1980: 7). In time, eleven rail lines were constructed to align with the canals, connecting to existing rail networks throughout the country (Chandler 1972; Powell 1980: 13). This intensive capital investment infrastructure demanded that companies find ways to ensure uninterrupted or, even better, escalating loads of coal to fill railcars. As such, coal production within these corporations became subordinate to the financial security of transportation interests (Aurand 2003: 21–22). Coal-rich properties were strategically exploited in a manner that may have been disadvantageous or inefficient to coal operations.

The solution for many was the consolidation, integration, or otherwise monopolization of coal production by transportation interests. By the third quarter of the nineteenth century, this monopoly came to be known as the "anthracite combine," a "complex network of interlocking relations" between transportation and coal operators that conspired to control supply, demand, and commodity values. Production costs, translated into worker's wages, were frequently used as pawns in this struggle. Independent coal operators paid as much as 60 percent of profits from coal mining to this monopoly of transportation interests, requiring the cutting of production costs (Aurand 2003: 19).

A unique landscape reflecting the political economy of the coal industry began to take shape by the 1850s. Depending on the region, coal operations were run either by large transportation conglomerates, such as the Philadelphia and Reading Railroad and the Lehigh Coal and Navigation Company, or by smaller independent family-run operations, such as that of Lattimer Mines. The former variety tended to operate under a form of absentee ownership connected to a corporation located in Philadelphia or New York (Rose 1981: 65). In contrast, Lattimer was one of about ten operations run by the local Pardee Family around Hazleton (Foulke and Foulke 1979). These family-run operations resembled feudal systems of land tenure, overseeing large tracts of land dotted with isolated small mining operations, coupled with worker settlements "strewn, by the caprice of the worn and upended strata" (Berthoff 1965: 262). This particular arrangement of company town, or patch town, reflected a situation in which companies maintained ownership of practically everything within the boundaries of the landscape: parcels, houses, roads, retail businesses. In these isolated environments, workers became dependent upon company paternalism, often through the binding force of debt.

"Free towns" also developed in the region, providing services to the coal industry as well as rental housing for workers. In time these commercial centers, although tied to the interests of the coal industry, developed diversified economies producing a middle class of professionals and merchants (Aurand 2003: 9). In the north, Scranton and Wilkes-Barre became the largest cities in the region. In lower Luzerne County, the city of Hazleton developed after 1820 and was incorporated by 1840. Between 1850 and 1890 its population grew from 1,500 to 12,000 people (Rose 1981: 66).

With the completion of transportation networks and the technologies for domestic and industrial consumption distributed throughout the tidewater market, demand for anthracite rapidly increased. Between 1822 and 1842 consumption jumped from 2.5 percent to 86.5 percent of total coal usage (Chandler 1972: 154). Investment in the region rose precipitously every year between 1820 and 1850 as small and large companies proliferated. From a total production of 365 tons in 1820, it exceeded 3 million by 1850. This massive expansion flooded eastern markets, such that prices dropped by more than half during the same period (Powell 1980: 12). By its peak in 1917, the industry would produce a total of 100 million tons of coal (Powell 1980: 12). The pattern of overcapitalization and overproduction, coupled with an unpredictable consumer market, affected the industry throughout the nineteenth century, as labor and capital struggled to devise a stable relationship between wages, profit, investment, and market values (Aurand 2003).

Work in the Anthracite Industry

If the dependable and intense energy produced by coal set off an industrial revolution throughout the country characterized by mechanization, de-skilling, and the "veneer of steel, glass and light," the experience of laborers in the coal industry could not have been any more different (Johnson 2010: 266). The geological positioning of anthracite coal, deeply buried in narrow and sharply contorted seams, was infamously difficult to mechanize until the proliferation of strip mining machinery in the 1930s (Blatz 1994: 14; Jerome 1934: 121). In contrast to the technological and organizational adjustments confronting industrial workers in other industries, coal miners used essentially the same methods, tools, and skills for the entire first century of the industry (Arnold 2014: 3–4; Dublin and Licht 2005: 20).

Sandwiched between sheets of rock in strata of various thicknesses and declination, each coal seam involved some combination of caprice, foreknowledge, and risk to excavate. Miners chased anthracite seams across geological folding and faulting that defied the logic of aboveground space. The typical method for the excavation of anthracite was "the room and pillar" method. Working alongside laborers who did the lifting and shoveling, miners excavated a rectangular room, known in industrial parlance as a "breast," off the main stretch of tunnel, or "gangway." Between each breast, miners would leave a pillar of coal to hold up the roof, which was generally composed of slate. Miners excavated into the "coal face" in whatever direction or declination the coal seam traveled until it ended, blasting away coal with dynamite to loosen material (Blatz 1994: 12–13). Laborers then loaded coal into waiting carts to be pulled to the surface by mules, to waiting elevators, or to inclined slopes where the material could roll to a stop for processing.

As superficial deposits of coal were depleted in the early stages of the industry, the excavation of deep seams of anthracite grew increasingly difficult and costly. At an industrial scale, coal operations required major capital investment, including the minimal measures taken to sustain the lives of laborers working underground. Mere *survival,* not comfort, was the goal of such technologies. Industrial-scale mining operations required the technological means to ventilate poisonous and explosive gases and to drain water before humans could survive the subterranean environments as deep as 2,000 feet below the surface (Dublin and Licht 2005: 20–21). This required implementing an assortment of pumps, boilers, fans, and other ventilation apparatuses, as well as elevators to deliver product, workers, and tools in and out of the ground.

Miners in the region were paid through a contract connected to individual productivity, which could vary by region, company, working condition, or labor situation. Wages were tied, at different times and places, to carts filled, footage excavated, or the weight of coal mined. The size of the carts varied by company or mine. A company-appointed inspector frequently judged the wage-value of each cart, determining the coal's purity, and docking pay if it contained a large fraction of sediment or matrix (Blatz 1994: 14). Contract miners were obligated to pay for their own equipment and the wages of their laborers. As a result, miners were compelled to self-discipline their own productivity. This sometimes meant bypassing safety procedures to ensure the fastest productivity with the least expense. Even then, as informants in the region have reported, miners might receive "the snake" at the bottom of their paychecks, indicating a debit or zero-sum gain for their labor rendered minus expenses.

For underground workers, the everyday violence of dangerous working conditions sustained a constant threat to their mortality and health. Engels identified the working conditions of English miners that ring true across space and time when he wrote: "In the whole British Empire there is no occupation in which a man may meet his end in so many diverse ways as in this one. The coal-mine is the scene of a multitude of the most terrifying calamities" (Engels 1969: 274). During the nineteenth century the anthracite industry held the second-highest mortality rate among American professions, just below railroad work (Aldrich 1997: 361; Aurand 1985: 227). The Pennsylvania anthracite region led the world's coal mining regions in fatal accidents per 1,000 employees for the end of the nineteenth century, averaging 3.25 individual deaths between 1891 and 1896 (Aurand 1985: 228). Between 1870 and 1913, the peak of industry, anthracite mining killed an average of 400 men a year (Aldrich 1997: 361). Even after a series of safety regulations were passed by the Pennsylvania legislature beginning in 1870, underground workers faced a number of threats ranging from explosive and poisonous gases, ceiling collapse, drowning, crushing by equipment, and kicking by mules. These accidents were frequently blamed on miners' refusal to respond to safety measures. Aurand (1985) suggests that miners rarely disputed this fact, but rejoined that carelessness came out of the economic necessity for speed (Aurand 1985: 232). In the last two decades of the nineteenth century, coal operators, politicians, and outside observers blamed the influx of new immigrant workers for the supposed increase in accident rates (Aldrich 1997: 366). In fact, among other factors, risk increased as upper strata resources were depleted, and increased mining depth multiplied the chances of roof collapse, flooding, and gas explosion. Despite this, fatality rates for underground workers improved in this period, possibly owing to the passing of safety laws and increasing enforcement (Aldrich 1997: 366).

Mining operations involved the labor of an army of workers other than miners, both inside and outside the subterranean realm conducting day to day operations (Blatz 1994: 12). In contrast to the bituminous coal industry, the complexity of mining anthracite required a considerably larger workforce to process the coal following its removal from the face. In 1889 only 30 percent of anthracite colliery workers worked in underground extraction, in comparison with 80 percent in the bituminous industry. The remainder of the workers at anthracite collieries typically worked aboveground maintaining operations or transporting and processing coal (Blatz 1994: 12). Outside labor included an army of engineers, inspectors, laborers, carpenters, blacksmiths, mechanics, machine operators, breaker boys, and administrative staff (Miller and Sharpless 1998: 83–134).

Laborers ranged in age from as young as six until workers were considered no longer employable. In 1870 the Pennsylvania Legislature enacted a safety code for the industry that barred children below the age of twelve from working underground, but it often went unenforced (Aurand 1985: 227). Labor tasks were seriated by age, a system that perennially reproduced the core of the workforce as children worked their way up the labor hierarchy (Powell 1980: 15). Children were employed to drive mules, to control the many doors inside the mines, to stop coal carts, and to assist in nearly every task outside the mines. Most infamously, children were employed in the coal breakers (Powell 1980: 16). Coal frequently came to the surface mixed with the surrounding geological matrix of slate and other materials. During the first century or so of anthracite mining, children and aging laborers crushed, sorted, and washed this material in the impressive structures known as breakers that once dotted the landscape. A newspaper article from 1897 describes an accident at the No. 3 breaker at Lattimer. Note the practically ubiquitous citation of personal blame for the accident.

> A sad accident occurred at the No. 3 Lattimer breaker yesterday in which Wm. Linderman of Harleigh lost his life. The boy was employed in the capacity of conveyor or tender, his duty being to see that rice coal, which is conveyed to the boiler house by this line, was equally distributed in the various bins. . . . An Italian who is employed about the breaker noticed the little fellow making an attempt to cross between conveyor or chain and told him he had better not do so. The boy did not heed the advice given and a moment later he was crushed to death. The Italian at once gave

> the alarm and the machinery was stopped but too late to save the victim as death was almost instantaneous. The boy was 15 years of age and was almost sole support of a widowed mother and several children, the father having died some years ago. A boy who has lately reached the age of 12 is employed in the same breaker where the accident occurred. ("Death in a Breaker," *Wilkes-Barre Times Leader*, July 27, 1897: 3)

After about 1900, technology and market factors, including mechanical sorting and other automated means, reduced the utility of child labor and the use of mules in the mines. By the time the New Deal brought secure child labor laws, technology had been developed to replace those tasks traditionally given to children. For this reason, Aurand argues it was not morality or law but technology and economics that first motivated the removal of child labor from the industry (Aurand 2003: 78–79).

LABOR AND IMMIGRATION (1860–1917)

The historic development of power and wealth in the region produced an environment ripe for aggravated class and racial tensions. High investment costs, capricious markets, natural disasters, monopolistic regulations, overproduction, and labor unrest all challenged coal operators' capacity to produce steady profit from anthracite extraction. To make up for this risk, they took a variety of measures to ensure that costs would remain lowest where they perceived the most room for flexibility: worker's wages. One tool they used to maintain these low wages was a competing army of surplus labor. The transnational flow of proletarian immigrant labor throughout the nineteenth and twentieth centuries coincided with the need for labor in the anthracite industry. These stories are inseparable.

Between the late eighteenth and mid-nineteenth centuries, systems of serfdom in Europe came to an end with the emancipation of the peasantry. As the scale and practice of agriculture moved toward consolidation, land became scarce as wealthy landowners bought up available acreage and displaced tenants. The resulting productivity caused global prices of many products to drop. Small-scale agricultural producers could no longer survive on meager scale economies without sufficient cash. These new systems of land tenure freed the peasantry of the obligations, but also the protections, of the feudal system. As mentioned above, Marx termed these preindustrial populations *vogelfrei*, literally "the bird-free," who wandered freely with only their labor as a commodifiable property. Simultaneously, manufacturing was becoming centralized and increasingly mechanized in urban areas, raising demand for the unskilled labor of the mobile peasantry (Hoerder 2002: 333). Moving to urbanized areas, and sometimes across the ocean, these peasants and their families made up the proletarian mass migrations between the 1820s and the turn of the century (Hoerder 2002: 331). North America received about a third of this migration, which, between 1815 and 1939, amounted to "50–55 million Europeans—or about one-fifth of Europe's entire population in 1800" (Hoerder 2002: 331–332). These changes swept across first Western, then Eastern and Southern Europe, developing systems of migration both circular and unidirectional.

The demographics of the anthracite region reflected this human mobility throughout its history. Beginning in the early stages of the industry, companies influenced the character and flow of immigration to the region, though this was far from the only way the region was populated. Company agents were sent overseas and to major ports courting new immigrant groups to join the Pennsylvania anthracite labor pool (Barendse 1981: 7–8, 24–28; Brooks 1898; Greene 1968; Roberts 1904). The first arrivals were skilled miners from Scotland, Wales, England, and Germany, who introduced craft techniques for anthracite's difficult extraction. (Before this time, the British Isles contained the world's only major anthracite mines.) Itter (1934: 31) reports an article appearing in the *Pottsville Weekly Miner's journal* from the summer of 1827 detailing the arrival of sixteen experienced miners from England, brought to the region along with their tools of trade to introduce the techniques of excavating anthracite. Reportedly, a fund of $1,000 was raised by industries to pay for the transport of miners to the region. The paper editorialized: "It is high time our mining operations were reduced to something like a methodical system of working. We have been bungling, making shift, and skimming the surface of our coal fields long enough" (*Pottsville Weekly Miner's Journal*, June 16, 1827, quoted in Itter 1934: 31). Bringing with them the skills and minimal resources needed to establish a social position in the region, these early migrants would frequently either leave the region for other work or ascend to managerial or administrative positions in the labor hierarchy (Rose 1981: 64–65).

Irish workers and their families migrated to the region between the mid-1840s and 1880s. The first wave fled the Potato Famine beginning in 1845, though emigration continued as English landowners evicted tenants to transform land usage from tillage to pasture (Miller & Sharpless 1998: 138). These Irish immigrants brought with them the class disadvantages of poverty brought on by exile or a lack of a competitive skill base in mining (Berthoff 1965; Dubofsky 1996). In a pattern echoed and elaborated upon in the region throughout the end of the century, colonialist antipathies based upon religious and ethnic divisions were recapitulated in the anthracite region, interwoven with work-related tensions (Itter 1934: 32).

During the period of the Civil War, demand for coal increased dramatically (Keil and Usui 1992; Palladino 1990). In some instances, miners took advantage of positive industry conditions to make demands of the industry (Aurand 1991: 301). The long history of unjust social and material conditions, coupled with growing class consciousness, led miners to rebuff entreaties to cooperate with the war effort, much less to contribute men to the fighting. Draft riots divided the region, particularly in the southern fields (Palladino 1990). Nonetheless, some miners did serve, including a large contingent of men from Hazleton (DeFazio 2008).

Coal operators, closely connected to war interests throughout the industrial northeast, increasingly put down strikes with the confidence and power of state support during and after the war (Aurand 1991: 301; Palladino 1990). In an action redolent of twentieth century colonialist intervention, a contingent of federal soldiers occupied the Southern coal field throughout the war at the invitation of coal operators, suppressing labor and draft resistance. Appealing to the industry's patriotic connections to the war, operators associated striking behavior with treason to garner federal support for suppressing labor radicalism (Aurand 1991: 301; Palladino 1990).

By the end of the Civil War, definitions of race and citizenship at the center of the conflict remained essentially unresolved despite the passing of the 14th Amendment extending citizenship to many disenfranchised groups. In the anthracite region, nativist patriotism, integrating anti-Catholicism, craft unionist protectionism and anti-immigrant sentiment spread during the period after the Civil War, intensifying class and ethnic tensions. Infamously, violence erupted in the region in the 1870s when Irish miners killed several Welsh and English superiors under the auspices of a secret society known as the Molly Maguires (Dubofsky 1996:40–45; Miller and Sharpless 1998:136–170). This initiated a pattern which continued throughout the nineteenth and early twentieth centuries as an accretion of immigrant groups encountered each other in the social order of the region.

Beginning in the 1870s, the largest number of immigrants came from Italy and Eastern Europe including ethnic groups such as Poles, Slovaks, Galicians, Tyroleans, Lithuanians, Ruthenians, and Hungarians. These groups made up what was termed the "New Immigration." In the largely agricultural economy in southern Italy, a regional depression caused a competitive demand for cash to purchase land released by the collapse of feudal land arrangements. Italians responded by emigrating throughout the world (Gabaccia & Ottanelli 2001; Hoerder 2002: 341–342). Some ultimately returned to purchase land, though many stayed in their new countries. About 18 million Italians left the country between 1876 and 1930. A third of these migrants moved to North America (Hoerder 2002: 341–342; Ngai 2004a). The contemporary wave of migrants from Eastern Europe included many nationless subjects of the Russian, Austro-Hungarian or German empires (Dillingham Commission 1911:661; Ngai 2004a). Besides the economic pressures from changing land practices, many of these populations suffered under political, religious and ethnic repression (Ngai 2004a: 120).

Immigrants from Italy and Eastern Europe outpaced that of earlier migration networks by the end of the nineteenth century, settling into labor hierarchies much as previous generations had done. Roberts (1904) reports the percentage of "foreign born" in the three counties of Lackawanna, Luzerne and Schuylkill, including immigrants from both the "new" and "old" immigrations, changed little between 1860 and 1900, ranging from 27.70 percent in 1860 to a high of 30.84 percent in 1870. The following chart, reproduced from Robert's (1904:19) *Anthracite Coal Communities,* delineates the percentage of increase in migrants from regions of origin between 1861 and 1901. Of those laboring in the mining industry, Roberts estimates that by 1900, 68.8 percent

Table 2.1. Immigrants to the anthracite region between 1861 and 1901

Years	Slavic and Italian	British and German
1861–70	1.05%	77.38%
1871–80	6.44%	57.46%
1881–90	17.65%	52.72%
1901	68.50%	13.50%

Roberts (1904).

were foreign born, about half of which are Eastern European or Italian (Roberts 1904: 20, 22–23).

Capital interests benefitted from, and even purposely exasperated, the destabilizing competition that developed between ethnic groups and longer-established generations of labor in the Anthracite region. Through a variety of factors, racialized nationalities of each group occupied places in a hierarchy, leveraging within and against their capacities to resist exploitation. Companies reified and exacerbated the discourse of racial hierarchy through wage practices, discriminatory litigation and the institutionalization and reinforcement of particular spatial and material conditions (see Bethoff 1965; Mulrooney 1989; Novak 1996).

Margaret Mulrooney, in her work on pluralist coal mining company towns of the bituminous fields of southwestern Pennsylvania, charts the feedback loop whereby ethnicity and race clearly define the limits to movement:

> the ethnic group to which a miner belonged determined the status of the job he held; this, in turn, determined his earning power. Companies then used earning power to compute the amount of rent an employee could afford, generally one-fourth of his monthly wages. Each company then used that figure to calculate how much it would spend on construction so that, ultimately, the amenities provided were a direct result of how much the employee earned. And to bring the cycle to a close, companies based the exclusion of provision of certain amenities on the ethnicity of their workers. But unlike other cause-and-effect relationships, this progression was not proportional; that is, an employee might alter the status of his occupation, and hence his earning power, but he

> could never change his ethnic origins and so remained somewhat limited in terms of housing. (Mulrooney 1989: 135)

As Mulrooney suggests, racialized identity often set the material limits for movement within the economic sphere in ways that advantaged the owners of the coal mining operations. It is likely that a similar system prevailed in the company towns throughout the anthracite region, which served as precursor and model to company town landscape and policies of southwest Pennsylvania.

Contemporary observers describe the process of "displacement" of older groups arising from a number of factors related to the "pull of industrial and social ambition and the push of racial friction" (Dillingham Commission 1911b: 661; Roberts 1904; Warne 1904). In 1904, Roberts (1904: 17–18) reports on the tendency of older generations to leave the region as soon as possible:

> The elements of our population have been, during the last thirty years, in a condition of flux. . . . Mining is disagreeable work. The thrifty members of our society rise to more congenial employments. Hence, annually, there is a withdrawal and migration of the "elite" of the group. . . . As soon as the receptive members of mining towns are trained in and acquire the principles upon which the social fabric of our institutions rests, they withdraw to the cities or migrate. (Roberts 1904: 17–18)

On the other hand, racial antipathy is cited by the Dillingham commission (1911b: 661) as the central cause of this social ambition when it reports on the older generation's "distaste for mine work *since* the immigrants entered it. . . . inducing the English-speaking miners to change their occupations, and . . . preventing them from allowing their children to enter the industry." Domestic occupations are also vacated by "ancient householders" when immigrant households or businesses arrive nearby, disturbed by the "nocturnal hammering" of household occupations or the "vociferous joviality" of a saloon (Dillingham Commission 1911b: 662).

Migrants in the coal industry were the subjects of classificatory work in the Dillingham Commission, a federally sponsored fact-finding effort to inform lawmaker's decisions on how to structure immigration restriction. Employing social science and scientific racism, the research tied the economic and social contributions of immigrant labor to their potential contributions as future citizens (Ngai 1999; Pula 1980). In a government report on the bituminous coal region from 1911, racialized aptitude and nationalities of the pluralist workforce are compared in terms of "Relative Efficiency," "Industriousness and Efficiency," "Adaptability and Supervision," and "Sobriety." This section concludes with a chart neatly matching the hierarchically arranged chaîne opératoire of the extraction process with the preferred racial groups for each task (Dillingham Commission 1911b: 549–550). By equating race, nation and aptitude, the solidity of these boundaries and their consequent divisions in quality of life, salary, work and living conditions were justified with a basis in natural structure.

A case study of one town in the anthracite region appraises the overall effects of the "social and moral deterioration" of the community by new immigrants through the vectors of criminality, health, sanitation and charity (Dillingham Commission 1911). Using an idiom borrowed from recent innovations in the science of sanitation and germ theory, the document describes immigration as an "infusion of a large element of foreign blood" of which "the two principal forces of its evil effects" include both "the conditions due directly to the peculiarities of the foreign body itself" and "those which arise from the reactions upon each other of two nonhomogenous social elements-the native and the alien classes" (Dillingham Commission 1911a: 671–672). The long list of "evil effects" includes items such as:

> "The lowering of the average intelligence, restraint, sensitivity, orderliness and efficiency of the community."
> "An increase of intemperance and. . . . crime."
> "An increase of sexual immorality."
> "A general loosening of the forces of social cohesion."
> "A civic demoralization of the ruling class."
> "A coarsening of the fiber of native-born through contact with the immodesties of the immigrant."

The deleterious effects of immigrants originate both from within the nature of the foreigners themselves and as a consequence of their interactions with older generations of Americans for whom a degraded environment is produced. The authors list the "most obvious" result of racial mixture as the "incapacity of local government and the wasteful administration of public funds" evidenced in the poor condition of infrastructure and municipal services (Dillingham Commission 1911a: 672–673).

EARLY EFFORTS AT UNIONIZING MINEWORKERS IN THE ANTHRACITE REGION (1868–1897)

The earliest effort to create a consolidated industrial labor union came in 1868 with the formation of the Workingmen's Benevolent Association (WBA). Organized by a mineworker from the town of St. Clair in the Southern field by the name of John Siney, the WBA would fall by 1875, victim to the age old issues of regional differences and militant operator resistance (Greene 1968:61–65). Throughout the remainder of the nineteenth and early twentieth centuries, the Anthracite Combination worked strategically to either appease or ruthlessly crush organized attempts to share in profits (Aurand 1968, 1991; Green 1969). This was accomplished with the support of law and order at the level of the state or federal government. For example, when a labor "riot" erupted in Hazleton in April of 1875, the governor of the State of Pennsylvania sent a company of militia into the region to pacify the strike (Aurand 1991: 304). By several accounts, operators spent lavishly on expanding and strengthening their private military police and private investigators during the strikes of 1875, infiltrating unions and protecting strikebreakers (Aurand 1991: 304; Greene 1968: 68). Though a number of regional or wildcat strikes erupted throughout the remainder of the century, it would not be until 1903 that the UMWA would once again organize the entire region (Aurand 1968; Greene 1968; Schlegel 1946).

THE EARLY HISTORY OF THE LATTIMER MINES

The Lattimer mines were first developed by the Pardee family and associates in 1865 (Foulke & Foulke 1979: 131–132). By this time, A. Pardee & Co. was the largest coal operator in the Lehigh or Eastern Middle coal field, operating a number of mines scattered around Hazleton (Foulke and Foulke 1979: 67; Greene 1968: 8). Generally, coal businesses in this field were characterized by independent family ventures with shared capital reinforced by nepotism and filial obligation.

These family dynasties were led by a cohort described as "petty coal kings" (Aurand 1968: 170) with a network extending to a variety of businesses, banks, and community interests (Foulke and Foulke 1979: 95). Ario Pardee, the patriarch of the family, made his fortune by the 1850s through the establishment of the major mines in Hazleton. In 1852 Ario built the Pardee Mansion in the center of town, surrounded by "spacious lawns protected by a high iron picket fence, broken at regular intervals by thick stone posts" (Foulke and Foulke 1979: 39). As suggested by the biographers of Calvin Pardee, son of Ario and owner of the Lattimer Mines at the time of the Massacre, independent entrepreneurial families such as the Pardees held any collective efforts to affect their business as "counter to [a] deep-seated American tradition" of independence and self-reliance (Foulke and Foulke 1979: 72). For the independent operators these efforts included both the increasingly powerful industrial labor unions and the combination railroad trusts which competed to siphon profits. Between the two, the biographers suggest, Ario considered labor unions the worse of two evils.

During the strikes of 1887 and 1888, the independent operators of the Middle coal field particularly resisted what they saw as an infringement upon the freedom of their independent financial sovereignty. Ario Pardee, representing the independent operators including Markel and Coxe, stated that, "Our position always has been, and is now, that we are unwilling to negotiate with anyone outside our employ, who knows nothing of our business and who is in no way connected with us; we are just as firm in that position as we have ever been" (quoted in Aurand 1968: 175; Foulke and Foulke 1979: 76). Along with the other independent operators, Pardee restricted credit at company stores to starve out the striking miners. A song written by the miners at this time celebrated the possibility of intergenerational class solidarity:

In looking o'er the papers now
A funny thing appears
Where Eckley Coxe and Pardee say
They'll stand for twenty years.
If God should call us miners off
We'll have children then alive
Who will follow in our footsteps
Keep the strike for thirty-five. (Foulke and Foulke 1979: 76)

By some accounts, when the miners struck, Pardee reportedly hired Italian strikebreakers to keep his mines running (Foulke and Foulke 1979: 76). However, some suggest Ario and his dynastic legacy harbored great suspicion and fear of new immigrant elements owing to the fact that they "not only interfere[e] with [Ario's] right as an American to run his own business, but even constituted a dangerous threat by foreigners to the established American way of life" (Foulke and Foulke 1979: 73–74).

EPILOGUE

The politics of race and the economics of laboring classes are entangled forces which structure the social history of the Anthracite industry of northeast Pennsylvania. So too is the materiality of its landscape, with an isolated rugged environment which only unforgivingly yields the coal buried deep below. These forces clashed in cataclysmic ways in the buildup to the Lattimer strike of 1897 that resulted in the massacre of nineteen immigrant laborers. In the last two decades of the nineteenth century, even as the region suffered through a series of crippling depressions, new immigrants continued to fill the ranks of the army of surplus labor, replacing earlier generations of nativized migrants. When a series of natural disasters pushed the local economy even further towards the precipice of economic disaster, local elites began to act. Structuring their argument in the national discourse of anti-immigration, operators, organized labor, lawmaking bodies, and the general public began to turn on the new immigrants, blaming them for overproduction, decreasing wages, the degeneration of the region's social life and the general decline of local economy. This coalition turned to the tactics of legal means to legislate immigration into the region, passing a series of laws beginning in 1889 aimed at limiting the influx of new immigrant families. In 1889 a law was passed requiring miners to pass a law in English for state certification (Greene 1964:201).

Organized labor, which for many years had earned growing support from the public and the press, began to strengthen its legal ability to alter working conditions. John Fahy, a local UMWA union organizer, temporarily gave up organizing to pursue political means to better the position of the established working class. This pursuit included passing laws to protect earlier generations of organized labor from competition from recent immigrants. In 1896 he joined a three-person committee devoted to producing and lobbying the governor and legislature to pass prolabor legislation (Greene 1964: 202). In 1897 the governor expanded the requirements for miner's certification. With the support of Fahy he also produced another anti-immigrant measure, the Campbell Act. The strategy of the Campbell Act was targeted at employers, who were taxed an extra three-cents per day per immigrant employed. Naturally, this tax would be passed on to the laborers (Greene 1964, Novak 1996). In the Middle coal fields, where independent operators were less inclined to appease their workers by minimizing the effects of the tax, the deduction of wages greatly angered the new immigrants. In August of 1897, amidst a series of wildcat strikes throughout the region, the violent disciplining of a superintendent provided the flashpoint to radicalize new immigrant workers to the surprise and dismay of the independent operators. This tension would culminate in the event known as the Lattimer Massacre.

3

The Memory of Violence

The Lattimer Massacre of 1897

It was not a battle because they were not aggressive, nor were they defensive because they had no weapons of any kind and were simply shot down like so many worthless objects, each of the licensed life-takers trying to outdo the others in butchery.

The Hazleton Daily Standard, September 11, 1897

The Lattimer Massacre took place over about three minutes in September of 1897. A posse of as many as 150 deputized local men armed with pistols and rifles shot immigrant coal mine laborers marching on strike in the town of Lattimer, Pennsylvania. The posse killed at least nineteen and wounded as many as forty more, emptying their rifles into the backs of the fleeing men. The victims were Eastern Europeans, demonstrating for fair wages and working and living conditions (Novak 1996; Pinkowski 1950). Many details of the event are obscured by the passage of time and by many conflicting accounts. This chapter focuses on the context and occurrence of the event, offering an archaeological account which adds some level of specificity and materiality to its ambiguous narrative.

In such incidents of explicit subjective violence, we can see the crystallization of commonly hidden social antagonisms suddenly materialized. Through encounters between migrants and those who act as representatives of the law, we can come to understand how political bodies define themselves through and against the "other." The migrant laborers, in their vulnerability to law and extrajudicial practice, exemplify the precarious condition of all subjects of the state by illuminating the ideological gaps between political identity and bare human life (Agamben 1998; Kristeva 1991). In the context surrounding the Lattimer Massacre, we must be attentive to the intersecting discourses and materialities that served to justify the "licensed life-taking." For in such moments when life and death stand in the balance, we can see "the threshold on which violence passes over into law and law passes over into violence" (Agamben 1998: 32).

HISTORICAL CONTEXT OF THE MASSACRE

Tensions were on the rise between capital interests and labor in the years leading up to the Massacre of 1897. Beginning in the early 1880s, a series of global and regional economic depressions affected the industry. Leading up to the Panic of 1893, the price of coal dropped to the lowest it had been since its peak in about 1865, with the exception of the 1877 depression. In these uncertain times, coal operators had recourse to a few options to maintain profitability. They could moderate productivity to drive up prices by work stoppages or by limiting the coal cars available for miners (Marsh 1987: 340; Walker 1924). They could also cut production costs by mechanizing extraction processes and decreasing labor costs. In testimony to a federal investigation of these monopolistic price controls during the strike of 1888, an attorney described this system as one in which "[the coal operators] take first the miner by the throat with one hand and the consumer by the throat with the other" (U.S. HoR 1889: 530).

In this period, coal mine operators always maintained a pool of surplus labor. Doing so granted them the logistical flexibility to respond to the exigencies of an unpredictable market. In 1898 one operator admitted to a reporter that he "must have more men on hand than could be used at any except for the best business periods" (Brooks 1898: 306). Throughout the last few decades of the nineteenth century, immigrants from Eastern and Southern Europe filled these ranks, joining or replacing earlier generations of workers.

As in many industrial contexts, these "new immigrants" were at first put to work in relatively unskilled laboring roles, and paid considerably less than longer established groups (Barrett and Roediger 1997; Dubofsky 1996; Roediger and Esch 2012). At this time, coal operators introduced mechanized processes such as strip mining and washery operations. Washery operations used mechanical means to reclaim a small fraction of coal from the ubiquitous waste, or culm banks, of the region. At their introduction, long-established underground miners recognized these new processes as efforts to bypass or destabilize the value of their craft skills. A motto used in strikes during the time incited a boycott of coal produced through the washery process: "Don't handle washery coal; that is what the operator stole from the miner" (quoted in Roberts 1901: 212).

The diversity of groups entering the region offered an additional advantage to operators, whose carefully balanced arrangements were threatened most by the possibility of organization. Competition, hierarchicalization, and language barriers made organization particularly difficult. Edward Pinkowski (1950: 209) alleges, "[Calvin Pardee] filled the houses at Lattimer largely with Italian immigrants and those in Harwood with Slovaks, Poles and Lithuanians. With a wholesale mixture of nationalities he felt that there would be less chance of a consolidation of the working men against his interests."[1] Despite their responsibility for the initial introduction of new immigrant labor to the region, coal mine operators nonetheless kept a wary eye out for signs of radicalism in their new workers.

Organized labor, lawmaking bodies, and the general public drew from national discourse to evaluate the social effects of the new immigrants. Organized labor, largely composed of nativized Western Europeans at the time, blamed the new immigrants for overproduction, decreasing wages, and the failure of strikes. They often cited a common belief of the time that the clannishness of new immigrant communal life would be an impediment to the development of class solidarities (Brooks 1898; Greene 1965: 11; Roberts 1901: 196–197).[2] By the last decade of the nineteenth century, organized labor had turned to politics to challenge these issues.

Legislating Anti-immigration in the Anthracite Region

By the 1890s, organizing efforts in the region had proven difficult for the United Mine Workers (UMW), the largest industrial union in the country. They turned toward politics, an avenue that had seen growing success nationwide (Dubofsky 1996: 74–82). In 1896, a new union representative by the name of John Fahy was appointed in the anthracite region. Fahy began his tenure by courting the new immigrants, offering speaking roles at rallies to Eastern Europeans as well as the nativized rank-and-file (Blatz 1994: 49). Despite these efforts, membership remained disappointingly low. In 1896, Fahy joined a three-person committee devoted to lobbying the governor and legislature to pass prolabor legislation (Blatz 1994: 53–54; Greene 1964: 202). In 1897, he helped to expand an 1889 labor law to include a language requirement for the licensing of underground miners, specifically targeting the prohibition of foreign laborers (Aurand 1985: 228; Greene 1964: 201). Furthermore, at Fahy's urging the governor produced a more explicit anti-immigrant measure known as the Campbell Act, which placed a tax of 3 cents per day on employers for each foreign-born male over 21 whom they employed (Greene 1964: 203, Blatz 1994: 54). In a UMWA journal editorial, Fahy praised the law:

> What a world of good this law would do to the American citizens who try to earn their living in the coal mines, if the tax were one dollar per day. I have an idea it would also do the foreign-born. . . . a power of good by keeping them out of the coal mines where all is cruel poverty and misery. (John Fahy 1896, quoted in Blatz 1994: 54)

Naturally, this tax would be passed on to the laborers (Greene 1964; Novak 1996). The courts and political establishment went a step further to close a possible loophole in this process. A few days before the Campbell Act was passed, the Naturalization Court of Luzerne County, which held jurisdiction over the greater Hazleton area, revised its rules, complicating the process for naturalization. The new rules required, among other steps, a petitioning process, the appointment of an attorney, and an examination on state and national laws in English. These procedural roadblocks greatly increased the cost, complexity, and time it would require to become a citizen, the only way to avoid the tax (Turner 1977: 16).

Racial thinking shaped the diagnosis of causes and projected solutions taken by the region's organized labor factions. The symbolic violence of racial discourse structured their considerations of the role of new immigrants in the industrial order. For a time, it distracted them from the culpability of the region's coal operators in slowing production, introducing competition, and de-skilling work.

The Anthracite Region of Pennsylvania: Summer of 1897

For the laborers of the region, work was sporadic throughout the decade of the 1890s. By the summer of 1897, collieries were open only two to three days per week (Novak 1996). Tensions were high in the region, with the Campbell Act passed at the beginning of the summer coupled with the extended stress of economic recession. Backlash from workers was temporarily neutralized in the Southern and Northern coal fields, where large corporations often pursued discretionary tactics to mollify their workers. In the Middle coalfield, where the independent operators interpreted these intercessions as acquiescence, this was not the case.

The precipitating event for the summer's strikes came in early August of 1897 at the Honey Brook Colliery south of Hazleton. A newly appointed superintendent by the name of Gomer Jones made logistical changes to save money. He specifically targeted the wages of mule drivers and the aboveground coal-stripping laborers, many of whom were recent immigrants (Blatz 1994: 56, 2002: 43). Jones' measures included removing some jobs from the payroll, lowering wage rates, and adding a tax to coal used by employees. He also consolidated the mule stables into one location, adding several hours of unpaid travel time to the mule driver's day. Presided over by a Slovak supported by an Italian assistant, the mule drivers struck on August 14th. They demanded the coal mine operators remunerate them for their travel time or that they reinstate the original locations of the mule barns (Greene 1968: 130). The strikers urged other workers to join them, setting up a picket line near the colliery. Brandishing an axe handle or crowbar, Gomer Jones confronted the strikers (Aurand 2002: 7; Novak 1996: 19–20). The exact sequence of events is not clear, but when a melee broke out, Jones struck a young mule driver by the name of John Bodan, breaking his arm. That evening, Bodan reported the assault to the police, naming witnesses among those present and providing the weapon as evidence (Novak 1996: 20). Word spread quickly about the assault and by the 16th of August many of the workers of the L&WB colliery struck in sympathy for the drivers. The striker elected two leaders for the opposition, a Slovak by the name of Józef Kinchila and an Italian named Nille Duse (Novak 1996: 27; Wolensky and Hastie 2013: 206). By August 20th the strikers came to an agreement with the company, winning some of their demands, including a promised analysis of wage rates from other collieries and an investigation of Jones' conduct. The strike resulted in an increase in union membership by new immigrant factions. John Fahy, back at work in the field, organized seven UMWA locals. Grouped by ethnicity, they amounted to about a thousand men in the region (Turner 1977: 25).

When the first wage deductions from the Campbell Act came into effect, the laborers unexpectedly found their paychecks diminished (Blatz 1994, 2002; Turner 1977). Sometime around the 25th of August of 1897, a strike wave hit the collieries belonging to independent operator A. H. Van Wickle. It was not the nativized underground miners who struck, but the young, mostly foreign-born slate pickers this time. The strikers demanded higher wages and an end to the system of company stores and obligatory company doctors (Blatz 1994, 2002; Greene 1968; Turner 1977). The strike quickly spread to other collieries, stretching into the beginning of September. Parades and rallies of Eastern European and Italian workers occupied the parks and streets of the city. Some wielded clubs and iron bars. They marched behind American flags to signal their belief that they were entitled to the rights they demanded as aspiring citizens.

On the 3rd of September 1897, the *Hazleton Sentinel* printed this description of a rally in McAdoo and the subsequent march toward Hazleton:

> By 10 o'clock this morning the vicinity of the hall on Blaine Street was black with people including many women. The Italian women are the most aggressive and they display a spirit that in such times is most dangerous to contend with. . . . The strikers gave the word to move and they came down to the corner, a solid mass of humanity that was formidable to behold. At the head of the column three men carried American flags and a large stoutly built woman carried a mallet. A score of young men had formed a line with their clubs held horizontally. Everything caught before this line had to either join the ranks or get out of the way.

On September 4, 1897, striking miners impatiently awaited the results of the eight-person committee charged with securing concessions from Head Superintendent Lawall of the L&WB company. The *Wilkes-Barre Times* (September 4, 1897; and Wolensky and Hastie 2013; Greene 1968) had this description, which, intentionally or otherwise, mocked the poor English of one of the strikers:

> It was a spirited meeting, full of Italian and Hungarian curses, threats and insinuations. . . . The committee reasoned, but of no avail, the miners were determined to give a demonstration. One burly Italian yelled at the top of his voice, "Whata da good of [the committee of] eighta da men! I'a kill a Lawall better alone!" . . . And to demonstrate that he meant what he said he drew forth a good sized carving knife and flourished it in the air yelling "Vendetta!" This burst of Italian eloquence tended to invigorate the crowd. . . . The Italian continued the matter in hand, "We getta do move on, and closa up the district," he said.

UMWA representatives took this opportunity to organize, forming locals and signing on a great number of laborers and miners. However, the newspaper reported that ultimately, their efforts to contain the energy and momentum of the strike was found wanting. A march beginning in the town of McAdoo on the morning of the 1st of September elicited a prophetic comment from a correspondent, who wrote: "The strong arm of the law cannot be subordinated to the designs of the inflammatory leader, and order must be maintained. Does it mean bloodshed? We hope not" (*Hazleton Sentinel,* September 2, 1897).

As colliery after colliery fell in the region to strikes, coal operators grew anxious. Determined to stop the lost profit and social instability, the coal company operators called upon the sheriffs of the three local counties to stop the strike. James Martin, Sheriff of Luzerne County, was called back from a vacation in Atlantic City to protect the areas in his jurisdiction. These included areas north of the city of Hazleton, including the town of Lattimer. On his return to the city, he met with coal industry administrative staff rather than with public officials. They made it clear to Martin that he would be held responsible if the mining operations in the region were affected. They also pledged that the companies would pay the expenses for the organizing and arming of a posse (Novak 1996: 90; Turner 1977: 28).

Martin deputized a posse of 87 local citizens, applying the law of *posse comitatus*. This law allows the drawing up of an armed force of deputized citizens when a situation of unrest such as a riot is present (Turner 1977: 28). Because of his unfamiliarity with the community, the local operators recommended Martin appoint two chief deputies chosen by them. They were two local businessmen with ties to the coal industry, Thomas Hall and A. E. Hess (Novak 1996: 119, 125). Despite the illegality of delegating authority, Martin tasked Hall with selecting and notifying the members of the posse (Coxe, in U.S. HoR 1901: 137). Nearly all the men involved had middle-class or professional positions in the city or surrounding regions. By name, all are of Western European origin.[3] Among them were half a dozen college graduates. Many had close connections to the coal industry including a banker, two civil engineers, a construction foreman, a mine superintendent, a salesman of blasting powder, a lumber merchant, bookkeeper, and coal company store manager. At least one coal miner, Alonzo Dodson, was also present (Novak 1996: 118–119, 124–126, 131, 158; Pinkowski 1950: 9). Twenty-three deputies were employees of Calvin Pardee & Company, twenty-two of whom lived in the town of Lattimer. Fourteen worked for the Lehigh Valley Coal Company. Six were members of the Coal & Iron Police (Turner 1977: 28).

The coal operators paid for all the expenses accrued by the deputies. The superintendent of the G. B. Markel and Company ordered rifles and shotguns, shipping them to a warehouse owned by the A. Pardee Company. Each member of the posse was given a Winchester repeating rifle and, by some accounts, a shotgun and/or revolver (Novak 1996: 90–92; Pinkowski 1950: 9–10; Turner 1977: 28–29). The shotguns were loaded with No. 8 shot (Pinkowski 1950: 9–10). The rifles were likely the Model 1895 Winchester repeating rifle loaded with .44–40 caliber bullets (Novak 1996: 136). In addition, the deputies were supplied with a trolley car so they could move unimpeded throughout the region. This vehicle, along with telegraphic communication, allowed law enforcement to travel in advance of the marching strikers (*Hazleton Sentinel*, September 3, 1897; Turner 1977).

A total of 500 deputies and 300 Pinkerton detectives joined other law enforcement to make up this force (Greene 1964: 206). For the next few days the deputies chased the strike across the town, attempting to neutralize its spread. For the most part, the striking laborers were orderly and quiet. Martin, known for his restraint, remarked to an acquaintance on the 9th of September that the posse was not entirely necessary given the fact that the strikers conducted themselves in a relatively orderly fashion. To a Mr. Frank Pardee on the evening of the 9th, Martin would complain that as the strikers were "doing nothing wrong," he therefore "had no right to interfere" with them (U.S. HoR 1901: 138). The coal operators, however, did not think this was the case.

By September 9th, the strikes shut down most of the collieries to the south of the city, putting great pressure on mine ownership. The collieries of Calvin Pardee to the north, including Lattimer, were an exception. Strikers shut down Harwood colliery, south of the city and occupied largely by Eastern Europeans, for a few days and organized a large UMWA local among its miners and laborers. They elected a Slavic laborer, Joseph Mehalto, as president and John Eagler, a Hungarian, as secretary. Upon the formation of the union, they sent a demand to the Pardee office for a raise of 10 cents in wages, a reduction in the cost of blasting powder, and the elimination of the company store and doctor (Pinkowski 1950: 11). The colliery at Lattimer, however, remained open. For the strike to work, production had to stop at all operations of the notoriously stubborn and powerful Pardee family business (Pinkowski 1950: 11; Turner 1977: 30).

SEPTEMBER 10, 1897: THE LATTIMER MASSACRE

On the 9th and 10th of September, representatives from Lattimer, first an Italian, and then a man by the name of John Glanati (alt. Glavati or Hlavaty), asked the men at Harwood to assist them with shutting down their colliery (Racek, in U.S. HoR 1901: 48; Novak 1996: 105, 109). The UMWA local declined to help but insisted that the marchers walk under an American flag and urged them to not bring any weapons. The strikers borrowed a flag from nearby Humboldt colliery (Eagler, in HoR 1901: 55). Approximately 250 to 300 men collected in Harwood around 1 o'clock in the afternoon and began the march to Lattimer, intending to pass through West Hazleton. On the way, they collected more men. Eyewitness accounts estimate the number of marchers ranging from 200 to 1,000 men, with most accounts estimating the size of the march as averaging between 200 and 300 men (Charles, in LFM 1898; Evans in LFM 1898; Hoyt, in U.S. HoR 1901: 85; Pinkowski 1950: 11, 12; Sherman, in U.S. HoR 1901: 50; Stiver, in U.S. HoR 1901: 51; Sykes in LFM 1898; Turner 1977: 30).

Law enforcement first stopped the marchers along their route into West Hazleton. Evan Jones, Chief of Police, and about forty of the deputies confronted the strikers, refusing them entrance into the city and demanding that they disband. A melee broke out and Deputy Thomas Hall broke a striker's arm (*N.Y. Times*, February 11, 1898). Deputy Ario Pardee Platt took offense at the strikers' wielding of the flag and took one of the flagpoles, broke it, tore the flag up, and threw it to the ground.

Following this first violent confrontation, the marchers continued toward Lattimer. The deputies made their way ahead by trolley to await them. John Welsh, a man from Hazleton who witnessed the encounter, spoke with Deputy Edward Turnbach in the trolley, who commented, "We have been marching around the country after these fellows for several days now. I don't see why the sheriff won't let us shoot some of them" (FLM 1898: 366). Other witnesses overheard phrases such as, "I'd like to get a pop at the sons of bitches. I'll bet I'll drop six," "Everyone of the Goddamn Hunks ought to be shot," and "We'll get even with them in Lattimer" (Miller and Sharpless 1998; Turner 1977: 31).

In Lattimer, about forty or more deputies and employees of the Calvin Pardee Company joined the posse, making up a total force of 80 to 150 men. What followed is classic example of the *Rashomon* effect in which the many divergent first-person accounts contradict each other. Only a schematic sequencing can be ascertained by sorting through these accounts. The sheriff and his posse first arranged themselves in a horseshoe shape running across the road into town. They then rearranged their lines so that they stretched from the edge of the road into a field to provide enfilading fire along the long axis of the column of marchers.

Deputy Hess later testified that the line of deputies was "75 to 100 feet from the road on the right and 15 feet on the left" (Hess, in U.S. HoR 1901: 141). Reportedly, when the marchers came into view, Sheriff Martin left the posse and walked toward the head of the line. He met the head of the line below a large gumberry tree, which locals would later memorialize as "the Massacre tree." After some discussion turned into struggle, someone fired a shot. The posse followed with a volley that continued for one to three minutes. The marchers turned and ran, but many got caught between the posse and the raised bed of the trolley line. Even as they scattered to the southeast and southwest away from the fire, the shooters continued to empty their magazines. One deputy pursued the retreating men, going so far as to run to the top of the steep embankment to fire after them, in the direction of a nearby schoolhouse several hundred yards away. The principal and his assistant witnessed the start of the event from the schoolhouse, and as strikers ran in their direction, the deputies turned to fire toward them, hitting the schoolhouse and felling several men yards away. In the chaos of the moment, the posse shot one of their own deputies in the arm (*Wilkes-Barre Times*, quoted in Novak 1996: 131).

Following the massacre, a shaken Sheriff Martin took the first train back to Wilkes-Barre. At the train station, Martin gave the first of several different accounts he would give of the moments before the firing began. To a reporter at the station he described how:

> They acted very viciously, reviling and kicking me, knocking me down and tramping upon me. I called upon my deputies to aid me, and they did so, but they were unable to accomplish much. I realized that something had to be done at once or I would be killed. I called to the deputies to discharge their firearms into the air over the heads of the strikers as it might probably frighten them. It was done at once, but it had no effect whatever on the infuriated foreigners, who used me so much rougher and became fiercer and fiercer, more like wild beasts than human beings. . . . I then called upon the deputies to defend themselves and shoot if they must to protect their lives or to protect the property that they had been sent to guard from being demolished. The next second there were a few scattered shots fired into the infuriated foreigners and a moment later the entire force of deputies discharged a solid volley into the crowd. I hated to give the command to shoot and it was with awful sorrow that I was compelled to do so, but I was there to do my duty and I did it as best I knew how. (Martin, quoted in Novak 1996: 143)

As Martin made this statement, recorded by a few reporters, and printed in the paper the next day, his attorney George Ferris took him by the arm and pulled him away to the hotel across the street. In the next public statement made by Martin he altered the story, perhaps by the coaching of his attorney. This time he reflected:

> When the strikers reached us I ordered them to halt and they did so. . . . A few of the men came from the head of the crowd and shouted that they did not care for me or my deputies and that they intended to go to Lattimer and stop the mine. While I was arguing with the men I saw them talking secretly and I knew that some trouble was brewing. The first thing I knew some big Italian came from the crowd, one of the men that had been shouting at me, and grabbed me by the throat. . . . The fellow pulled my head under his arms and struck me on the shoulders, and when I had a chance to look up I saw that I was surrounded by several fierce looking men. I shouted to the men that they should arrest the person who had attacked me, but in the confusion the deputies evidently did not hear what I said. Then I heard a shot and it was soon followed by another. This seemed a signal for a combined volley on the part of the deputies and before I could extricate myself from the crowd that had surrounded me there was one rifle crack after another. . . . The order to fire never came from my lips. When the deputies saw that I was attacked I suppose they thought it was their duty to protect me. (Martin, quoted in Novak 1996: 145)

By some accounts, Deputy A. E. Hess kicked a prostrate victim, continuing the violence even after the shooting ended. Upbraided for his behavior by a bystander, he replied, "Shut up or you will get the same dose" (Novak 1996: 131). Some deputies and bystanders placed the wounded on the trolley upon which they were transported to the Hazleton State Hospital. The conductor reportedly refused to carry the wounded because they could not pay fare until the schoolteacher gave him $1 to defray the costs (Marinko, in LFM 1898: 366).

The wounded overwhelmed the capacity of the hospital in Hazleton, requiring the expedited discharging of patients to make room (Pinkowski 1950: 17). Exact numbers of the wounded vary by account, but generally run between thirty-two and fifty (Miller and Sharpless 1998: 234; Pinkowski 1950;). Treatment included the amputation of limbs, perforation of skulls to relieve pressure from head wounds, and extraction of multiple bullets from abdominal wounds (Pinkowski 1950). In the *Philadelphia Inquirer* on the 14th of September, Dr. Kellar reported that, "The head cases are extremely puzzling. The bullets have imbedded deeply into the substance of the brain, which has oozed out through the openings of the skull. These bullets, of course, cannot be removed. Notwithstanding this the patients are conscious and spend portions of the day in conversation with their wives and families" (*Philadelphia Inquirer*, September 14, 1897). Meanwhile, some deputies fled town, hiding in a hotel in Atlantic City under assumed names (Miller and Sharpless 1998: 234; *Wilkes-Barre Times,* September 17, 1897).

RESPONSES TO THE MASSACRE

News of the event spread quickly throughout the region. Sheriff Martin, at the direction of Governor Daniel H. Hastings, called in the Pennsylvania National Guard under Brigadier General John P. Gobin (Miller and Sharpless 1998: 234). Five regiments of the third brigade mobilized, arriving by the morning of the 11th of September. General Gobin, fearing that the upcoming funerals might agitate the public to violence, "so distributed the [troops] that the crowds of foreign miners who attended these funerals should realize that in or near all quarters of the community for miles about there was an ample military force to preserve order" (Gobin, U.S. HoR 1901: 76).

The communities in and around Hazleton responded to the Massacre in several ways. The ethnic communities quickly made plans for the interment of the victims for the 12th and 13th of September. A funeral procession throughout the city preceded each burial, accompanied by mourners in ethnic regalia and music. Victor Greene (1968: 140) describes one "snake-like procession wending its way along the road":

> The day's weather supplied the perfect background for what took place. Through a fine, chilling drizzle and a heavy mist, a brass band playing the death march led the cortege. With muffled drums it preceded the three biers, the families of the deceased and a thousand paraders in grocery and beer wagons past culm piles crowded with fascinated onlookers.

General Gobin, in a meeting with religious leaders, "requested that in their sermons everything of a revengeful, or vindictive character or to provoke a breach of the peace should be omitted" (Gobin 1898). A brigade commander revoked an initial request by the National Guard that barred music from the funerals, in fear of the passion it might inspire (*Wilkes-Barre Times*, September 13, 1897). His demand that mourners could not display any weapons during the proceedings went unheeded as ten members of the Italian-American Society held their ceremonial sabers aloft at St. Stanislaus church (*Philadelphia Inquirer*, September 14, 1897).

At first the strike of the previous few weeks greatly expanded. Local communities held mass meetings in neighborhoods throughout the town to discuss how to proceed. The community of Lattimer, which had taken a back seat to the previous three weeks' events, became a hotbed of radical activity. Fifteen hundred workers at Lattimer walked off their jobs on the 12th of September (Beik 2002: 68; Miller and Sharpless 1998: 235). Strikers held a massive meeting in the town on the 14th and 15th. They advanced a number of demands for the coal companies, including the immediate dismissal of all officials who participated in the posse (*Philadelphia Inquirer*, September 14, 1897). A meeting on the morning of the 15th of September, attended by every member of the colliery, resulted in the demands for an increase in wages and the abolishment of the company store. Despite this unrest, Calvin Pardee, in an interview for the *Philadelphia Inquirer*, maintained that his men were "happy . . . until a mob forced them from work" (*Philadelphia Inquirer*, September 16, 1897).

In the aftermath of the Massacre, violence came largely in the form of rumors and threats. The *Philadelphia Inquirer* (September 14, 1897) reports that the miners in Lattimer planned to "blow up the officials and the company store with dynamite," although this rumor never surfaced. However, management assigned six extra watchmen to patrol the town. News articles indicate that divisions existed between the English-speaking laborers at Lattimer and the enclaves of Eastern and Southern European workers. An article in the *Philadelphia Inquirer* from the 15th of September reported that a committee of English-speaking miners formed to confer with miners "from the Hungarian and Italian colonies for the purpose of endeavoring to persuade the latter to be more reasonable in their demands" to the Pardee Company management. An article from September 18th shows that even stronger divisions between these two groups surfaced and that the presence of miners among the deputies proved to be a divisive issue in the town. The *Wilkes-Barre Times* (September 18, 1897) reported:

> At Lattimer . . . the forty or fifty English-speaking miners say they would go back at once if their safety is guaranteed but they say they dare not. . . . They say the foreigners think they are all deputies and that if they tried to resume work their lives would not be worth a moments purchase. The foreigners are quite determined not to go back and they are not at all backward in making threats as to what they will do if any of the others try to work.

On the 19th of September, the *Philadelphia Inquirer* reported that:

> A large party of Italians entered the home of Deputy Sheriff Jacob Dougherty, at Lattimer, early this morning. They searched the house from cellar to garret for the deputy whom they declared they had come to kill. A squad of soldiers from the Thirteenth Regiment was dispatched to the scene and arrived just in time to see the last Italian make his exit from the rear door.

A third article from the 15th of September reported that Italians were leaving the town *en masse* because of the shooting.

By September 16th, however, the extended strike took its toll on the miners and their families. As companies began to offer some amount of recompense, men began to return to work (Greene 1968: 142–143). To those holding out for more radical reforms, extending the strike was a critical matter. In the town of Lattimer and elsewhere, women were often at the forefront of efforts to push for more radical reforms. There are numerous accounts of cavalry troops confronting women armed with clubs, rolling pins, and frying pans (Greene 1968: 143–144; Pinkowski 1950). In Lattimer, "Big Mary" Septak, who managed a boardinghouse in town, led a group of women in attacks on any men who dared to return to work prematurely. "A short stout Hungarian woman," she "cavorted about in most amusing style," carrying a "crude sword fashioned out of a lathe" (*Wilkes-Barre Times*, September 20, 1897). An effort to shut down the washery at Lattimer by Mary Septak and her "amazons" was ultimately unsuccessful as troops pushed them off the road by the muzzles of their rifles (Pinkowski 1950: 27).

By the 20th of September, "the backbone of the great strike" was weakening as miners began to return to work (*Wilkes-Barre Times,* September 20, 1897). At Lattimer, 800 of the 1,500 workers returned to work, most of whom were Italian. The Eastern Europeans, on the other hand, noisily paraded through town the night before, rattling tin cans and chanting, demanding that the members of their community resist the urge to return. By October 4th, however, all the collieries in the region would be back in operation (Greene 1968: 144).

The national and international audience responded variously to the events of the Massacre. Often, they appraised the significance of the event in the context of broader political and social themes. The Austro-Hungarian government viewed the situation in international terms as an attack on the lives of their sovereign nationals. A prominent Hazletonian Hungarian by the name of John Nemeth brought the incident to the attention of the Austro-Hungarian consul in Philadelphia. They promptly began an investigation of the event, including a collection of eyewitness accounts from some of the marchers. They also assembled a list of confirmed citizens of the Austro-Hungarian Empire among the killed and wounded, noting age, birthplace, injury, and, for the dead, the number of surviving relatives. They promptly demanded a criminal investigation of the deputies, intending to force the federal government to supply indemnities to the surviving families of the fallen (Greene 1968; U.S. HoR 1901).

Other editorialists connected the tragedy to transnational events such as the concurrent U.S. involvement in the Spanish-American War. In the saber-rattling aftermath of the explosion on the *Maine,* a commentator invoked the massacre to question the nation's willingness to mobilize for war. He questioned why there was popular support for a war on a foreign power and yet for the "carnival of carnage that takes place everyday" in mines and industrial work places, "no popular uproar is heard" (Zinn 2003: 307). Similarly, a London-based paper suggested, "There is no reason for America to fight Spain after all. An outlet for her fighting energy is provided by the indiscreet vigor of a Pennsylvania Sheriff" (*Daily Mail,* September 14, 1897, quoted in Greene 1964: 210). An anarchist newspaper based in Chicago by the name of *Lucifer, Light Bringer* proposes a direct connection between the Massacre and rising American imperialism:

> If this is not imperialism, pray in what does imperialism consist? From the state of this centralized Federal power packing the Supreme Court with its creatures and partisans . . . the arrest of Emma Goldman and the massacre of the helpless strikers at Lattimer, Pennsylvania, this hydra-headed imperialism overshadows the land. In brief, the United States has finally become under the evolution of despotism. . . . Every avenue of the nation's life is fed and poisoned by a capitalistic corrupted and religion—by law—perverted public school system. . . . And worst of all, the enslaved and degenerate American, thanks to the public school, loves his chains. (Baylor 1898: 57)

Other media sources were supportive of the sheriff's indisputable duty to maintain the peace and the strength of law. On the 23rd of September 1897, *the New York Observer* wrote, "The first duty of government is to protect life and property, so to defend its citizens in possession of the results of their toil. . . . It is for such emergencies that sheriff's posses exist, and the law places no limit to the means they shall use."

THE TRIAL

By the middle of September, authorities faced mounting pressure to prosecute the sheriff and his deputies from the community, the Austro-Hungarian government, and an organization of citizens devoted to the cause known as the National Prosecuting and Charity Committee of the Lattimer Victims or the NPCCLV (Turner 1984: 25). Judges L. H. Bennett and John Lynch issued an arrest warrant for the sheriff and seventy-eight deputies on the 20th of September. The defendants were charged with killing and wounding the strikers "feloniously, willfully, and of their own malice" (Turner 1984: 25). The defendants were arraigned; a bail of $6,000 set for each, and a trial date scheduled for February 1, 1898.

The defendants comprised Sheriff Martin and sixty-eight deputies by the time the trial date arrived; charges were dropped for ten erroneously charged men. The court jointly charged the defendants with murder and a second count of manslaughter for each man killed. A separate indictment charged them for "the felonious wounding of each man wounded" (U.S. HoR 1901: 82). The initial trial was to be a test case, in which the defendants were tried only for the killing and manslaughter of one of the victims, Michael Ceslak. The remaining charges depended upon the results (U.S. HoR 1901: 82).

One hundred and three witnesses testified for the prosecution and 111 for the defense (U.S. HoR 1901: 83). The prosecution hoped to show that the posse was unnecessarily called and, by the time of the confrontation, already predisposed to violent confrontation (Greene 1968: 241). The case for the defense of the sheriff and the deputies depended upon evidence of the belligerence of the strike in the days preceding the march and during the confrontations at West Hazleton and Lattimer. Throughout the trial, many sources recounted the sequence of events leading up to the Massacre. Strikers, deputies, and bystanders gave eyewitness accounts of the incident, as did third-person observers such as public officials, lawyers, and media. Unfortunately, the trial transcripts mysteriously disappeared at some point in the early twentieth century, complicating the task of recreating the events. Official reports, newspaper articles, and dramatizations, however, recorded or summarized portions of the testimony.

A rich intersectional discourse about race, class, and citizenship runs through the rhetoric of both sides of the case. In his opening address, Martin's attorney, George Ferris, implied that the high standing of the professional men called to the posse should be evidence enough to counter the charge of wrongdoing. Ferris proposed that the sheriff "did not choose loafers, bar-room bummers, toughs and thugs but good, law abiding citizens—men of good judgment, good character, good standing in the community" (Ferris, in Palmer 1913).

Both sides invoked the flag as symbolic evidence of what was, purportedly, the higher stakes at issue in the trial. Speaking for the defense, an attorney "handled the evidenced flag the strikers carried," opining that, "they marched along the highway as labor has a right to march, believing that the flag they carried would protect them—not the red flag of anarchy, but the flag it is a pride to be wrapped in in death." (Palmer 1913: 113). For the defense, Attorney Lenahan (Palmer 1913: 116) responded:

> It is true this crime was done under one flag, that flag that was drenched in the blood of the patriots, that flag was to be the cover

> of the brigand who put himself beyond the pale of the law. It is for that flag you bared your breast to rebel bullets—to protect that flag. They would desecrate it!

Attorney Henry Palmer suggested that Lattimer could become "the place where was born a movement" to "wrest from the American people the liberty so dearly bought." Again, the image of the flag was evoked: "Men, unless throttled now, will rise up in rebellion against our government to tear down the Stars and Stripes and fix over the highest places the red flag of riot and confusion" (Palmer 1913: 119–120).

The trial went on for five weeks, and in the end the jury acquitted Sheriff Martin and his deputies of any crime. The attorney Lenahan's eloquent but theatrical address, couched in fear-mongering terms that played upon antiforeigner and antianarchist sentiments of the time, ultimately swung the jury despite overwhelming evidence. He described the strikers as "that lawless horde that came from the steppes of Asia [that] has found its way here." He continued, "The history of the Hun and Slav in the old country is that of mischief and destruction. And they marched under Attila ruthlessly over Europe. No home was too sacred or virgin too pure for their assault" (Lenahan, in Palmer 1913: 117–119).

ARCHAEOLOGY OF THE LATTIMER MASSACRE

In 2010, a team of archaeologists from the University of Maryland including Paul Shackel and myself sought to add an archaeological account of the Lattimer Massacre and, through this process, return the tragic event to the attention of the public. Members of BRAVO (Battlefield Restoration and Archaeological Volunteer Organization) of Monmouth, New Jersey, led by Dan Sivilich, aided our endeavor (Sivilich 2011: 1). Our goals for this survey were to assess the condition of the resource, determine the location of the Massacre site, identify the firing line, and locate the initial engagement. If the remains of the Massacre left at the site exhibited sufficient integrity, we also hoped to extrapolate the size of the posse and the variety of munitions employed, and to trace the movement of various agents across space.

First, we identified the approximate location of the Massacre near the west entrance into the town of Lattimer using archival sources and local memory. The property is typical of local mining landscapes, consisting of scrubby trees rooted in a matrix of coal-processing waste. A study of aerial photographs established that sometime after the Massacre and before 1938, the coal company deposited a large pile of coal-processing waste on the site. By the 1950s, much of this waste material was gone, leaving only a thick upper stratum of shale mixed with coal across the surface of the site area. The first house along Main Street, which figures in some accounts as a reference point for the location of the posse, had burned down late in the twentieth century and a new house built roughly in its location. The raised bed of the trolley tracks, which also featured so prominently in accounts, was also gone. A large tree along the road, known locally as "the Massacre tree" and which figures prominently in accounts as the location of the first violent interaction, had also reportedly been removed by municipal road maintenance in the years just before the survey.

The team spent two weekends in November and December of 2010, a total of about 113 hours, examining the site. Eight crewmembers employed metal detectors along ten-foot-interval transect lines laid through the underbrush. They followed this transect survey with a timed, semisystematic survey broken down into zones. We found a total of forty-three artifacts, including seven bullets of various calibers, twenty-two brass cartridges, six copper jackets, a miner's tin cup, a cupric metal suspender clip, a silver-gilded serving spoon, and several objects of unidentified copper alloy hardware (Sivilich 2011). The non–arms-related artifacts were most likely unrelated to the Massacre with the exception of the tin cup, which showed signs of perforation by shotgun blast.

Battlefield archaeologist Doug Scott aided in the analysis of munitions. Scott examined the bullets for tool marks left by firing pins, extractor and ejector marks, and barrel rifling. The sixteen shell casings all postdated the Massacre including some .30–06, .35 Remington, .30–30, and a single .32 pistol casing (Scott 2011: 2–3). Of the seven bullets found, four dated to the period of the Massacre. Those that decidedly postdated the Massacre were one .357 round, developed in 1935, and two .38 Specials, introduced in 1902 (Scott 2011: 3–4).

Crew found seven copper jackets clustered near the road where historic accounts suggested close-range firing occurred. Ammunition manufactures added copper jackets to bullets with the invention of smokeless powder. These rounds have an exposed tip called a soft point, advertised by Winchester as "metal-patched" and adopted around 1892 for the .30 Winchesters. The resultant increase in muzzle velocities from this innovation required that a jacket of copper alloy of 60 percent copper and 40 percent nickel be applied to bullets to prevent them from melting from contact with the barrel. During the Spanish-American War, these munitions were loaded into the new 1903 Springfield rifle, which exhibited an even faster muzzle velocity. The resulting heat tended to melt the cupro-nickel compound in the barrel (Hatcher 1966: 343–345). In 1922 ammunition developers solved this issue by adding tin and later zinc to the copper. We used a trace elements analysis, or XRF, to test the chemical signature of the alloy used in its manufacture.[4] A sample readout of one of the jackets showed peaks for copper and zinc content and the complete lack of nickel. This analysis revealed that these jackets also dated to after the Massacre, and likely after 1922.

Two of the bullets that matched the period of the Massacre were .38 caliber long and short rounds. These rounds were introduced in 1875 for the Colt revolver. Markings on them suggest they were fired from one or more Smith & Wesson revolvers (Scott 2011: 4) (Figure 3.1). A single, heavily impacted .22 round was also identified. Developed in the 1860s, the .22 is still the most popular small-bore round today. These three rounds were found in a cluster at the edge of the road, adjacent to the roughly plotted position of the "Massacre tree." A fourth round, a heavily patinated .32 caliber pistol round introduced in 1875, may also date to the Massacre. Marks on the bullet suggest it was fired from a Smith & Wesson revolver. It exhibited heavy patination akin to that of other period munitions identified. It was found in the middle of the field, approximately 170 feet from the gumberry tree.

The survey succeeded with some of the goals for the project, while others remain elusive. A cluster of three bullets established the probable location of the initial engagement and the "Massacre tree." The morphology of the bullets reveals some compelling information about that precipitous moment. An anomalous fourth bullet that may date to the incident was found at a location near where the line of deputies stood. (Figure 3.2) The survey recovered no spent cartridges dating to the Massacre. The absence of brass cartridges may reflect several scenarios. Locals may have collected them as mementos or souvenirs after

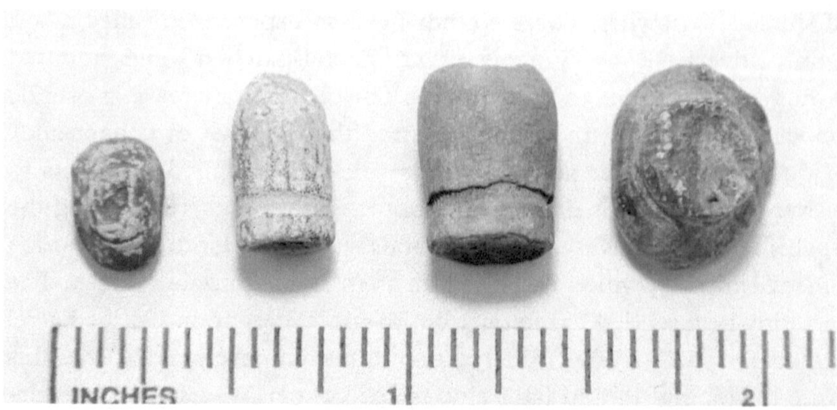

Figure 3.1. Bullets recovered from the site of the Lattimer Massacre. Artifact #s 258 (*L*) and 242 (*R*), .38 long and short revolver rounds (photo by author).

the event. Investigators may have collected them as evidence before the trial. Lastly, the cartridges may have been removed in the process of building up and removing coal-processing waste.

Of interest are the three bullets identified in the location of the initial engagement. All three were revolver bullets, two .38 Smith & Wessons and a .22, as opposed to the Winchester repeating rifles described in many accounts. If this cluster represents the initial volley, the archaeology has produced provocative evidence of who might have fired the first shots. When asked if the deputies were all armed with Winchesters during the trial, Martin asserted, "All except a few, who had buck shot guns" (Palmer 1913: 83). The archaeology raises the question of who of those present were armed with revolvers besides the sheriff. One possibility points to those previously armed with standard issue sidearms: the Coal & Iron police, Pardee Coal Company guards, or other law enforcement agents. As the responsible parties pressured with defending the colliery, law enforcement had an interest in ensuring that the posse did not capitulate with the strike. All historic accounts suggest that after the initial encounter, the firing continued for one to three minutes, as the deputies' emptied their rifles at the fleeing strikers as far as 300 yards away. This area lies across the road to the south of the area surveyed. Future research may return to survey this area as well.

The fourth bullet, a .32 revolver bullet, is an anomalous find that may have a number of explanations. It may be from an unrelated event

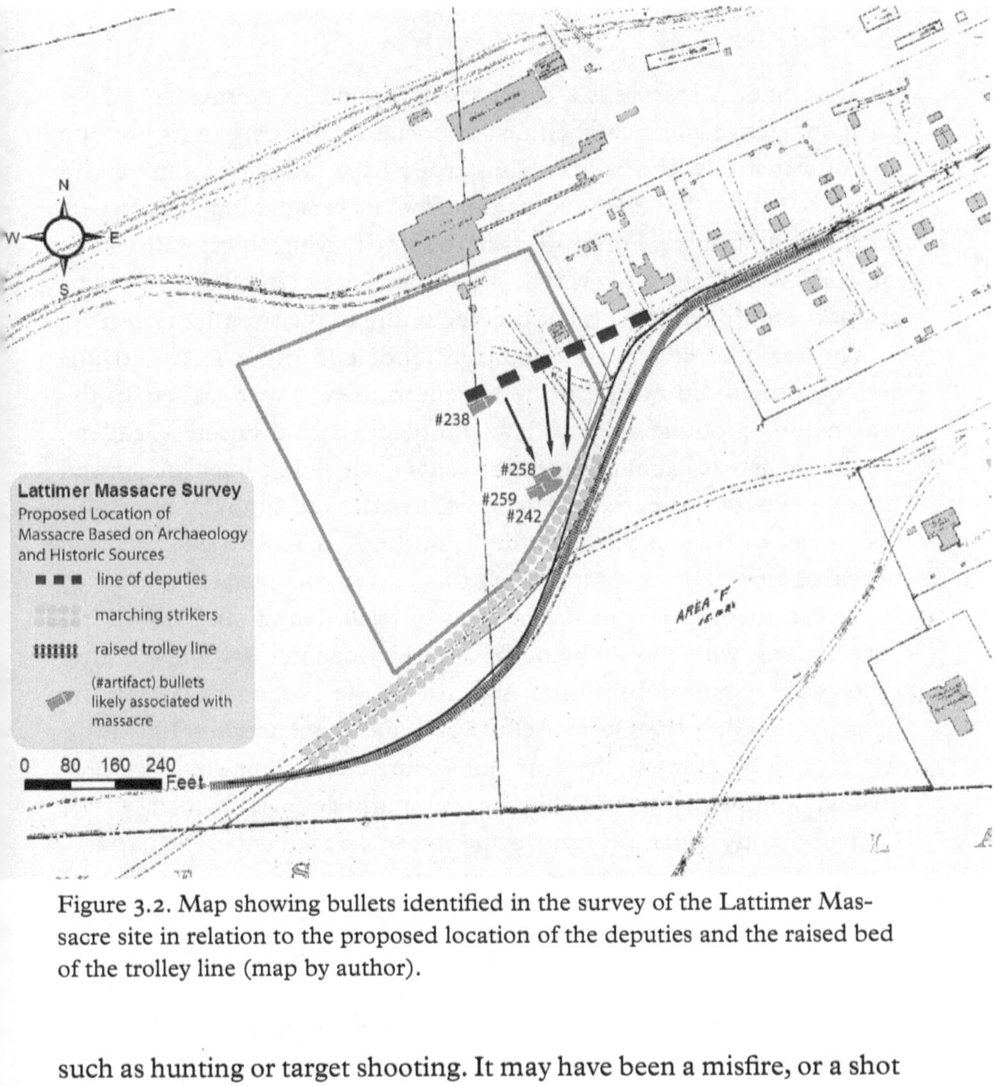

Figure 3.2. Map showing bullets identified in the survey of the Lattimer Massacre site in relation to the proposed location of the deputies and the raised bed of the trolley line (map by author).

such as hunting or target shooting. It may have been a misfire, or a shot fired into the ground by a reluctant or anxious member of the posse. Conversely, it might have been return fire from a striker with a pistol. While trial transcripts contain accounts of weapons among the strikers, they are controversial. In several instances, witnesses later admitted they were paid by the defense to relate this falsity in court (Novak 1996: 211–212; 215). Nonetheless, the archaeology presents us with the evidence for just such a possibility.

CONCLUSION

The Lattimer Massacre is a deeply scalar event. As an incident of explicit subjective violence, the most important details are in its buildup and aftermath. Of the brief encounter, perhaps little more can be said of its particulars that does not distract from its broader implications. Its greatest significance, I propose, is in the way in which it crystallizes the violence at the center of everyday life in this era of great social, political, and economic upheaval at the beginning of the twentieth century.

The regional efforts by organized labor and other forces to litigate migration did not occur in a vacuum. Newspaper readers of the era, including public officials, law enforcement, and business leaders, would see articles about domestic events such as Lattimer juxtaposed with news about imperial exploits, both written in similarly racialized discourse, such as in the example from the *New York Evening Journal* provided here (Figure 3.3). During this period, the racial character of American identity was in the process of legal definition. The federal government, with the lobbying of a variety of interests, passed laws such as the Chinese Exclusion Act (1882), the Foran Act (1885), the Scott Act (1888), the Geary Act (1892), and the Anarchist Exclusion Act (1902); in addition, the Supreme Court decided on Plessy vs. Ferguson (1896), and the the Dillingham Commission was formed (1911). Concomitantly, a parallel transformation was occurring in the nation's foreign policy with the Spanish-American War, the suppression and occupation of the Philippines, and the seizure and construction of the Panama Canal. All these events served as a backdrop to the domestic strife of events such as the Massacre. As a whole, this litigation and policy defined the ever-closer relationship between race and nation through discursive operations of justification (Greene 2009; Kramer 2006).

The bureaucratization of racial thinking during this time empowered all those called to execute these laws with the sovereignty of the state, defined as the power over life and death. Paradoxically, in binding them to the tautological power of the sovereign state, the laws made them "petty sovereigns" and unbound to law itself, as described by Judith Butler (2004: 56). At such a scale, the question of "who shot first?" or "who was acquitted in the trial?" becomes an abstraction: a Kafkaesque absurdity. An analysis of the racial dimensions of the

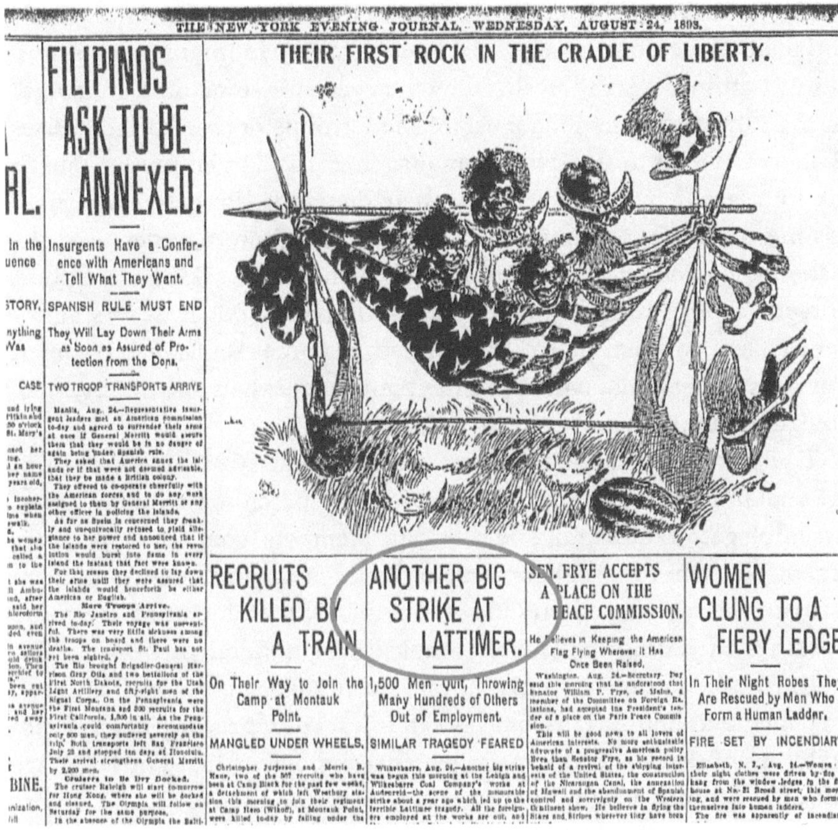

THE NEW YORK EVENING JOURNAL, WEDNESDAY, AUGUST 24, 1898.

FILIPINOS ASK TO BE ANNEXED.

Insurgents Have a Conference with Americans and Tell What They Want.

SPANISH RULE MUST END

They Will Lay Down Their Arms as Soon as Assured of Protection from the Dons.

TWO TROOP TRANSPORTS ARRIVE

THEIR FIRST ROCK IN THE CRADLE OF LIBERTY.

RECRUITS KILLED BY A TRAIN

On Their Way to Join the Camp at Montauk Point.

MANGLED UNDER WHEELS.

ANOTHER BIG STRIKE AT LATTIMER.

1,500 Men Quit, Throwing Many Hundreds of Others Out of Employment.

SIMILAR TRAGEDY FEARED

SEN. FRYE ACCEPTS A PLACE ON THE PEACE COMMISSION.

He Believes in Keeping the American Flag Flying Wherever it Has Once Been Raised.

WOMEN CLUNG TO A FIERY LEDGE

In Their Night Robes They Are Rescued by Men Who Form a Human Ladder.

FIRE SET BY INCENDIARY

Figure 3.3. Page from the *New York Evening Journal* showing news items reflecting American racialized imperial exploits juxtaposed with articles about the racialized labor violence of the Lattimer Massacre (*New York Evening Journal*, August 24, 1898, Library of Congress, Serial and Government Publications Division).

region's landscapes defined the *predisposition* for violence on the part of the deputies, just as it would preordain the outcome of the trial. In a sympathetic and revealing *Wilkes-Barre Times* editorial written the day after the Massacre, an author wrote, "It must be remembered that confronting the sheriff and his legal force was a desperate mob of unreasonable and unreasoning men whose carelessness of life both the sheriff and his deputies were thoroughly acquainted" (*Wilkes-Barre Times*, September 11, 1897). As the editorial might suggest, the eruption of violence might have seemed almost *predestined* at its occurrence to all those who witnessed it.

The event revealed significant rifts in the towns of Lattimer Nos. 1 and 2, breaking down along racial lines. The new immigrant communities of Lattimer were radicalized by the event for some time afterward. At the same time, the longer established groups organized themselves along racial lines to stand by the management and attempt to persuade the immigrant factions to lessen their demands. Perhaps more than anyone present at the time, the wives of coal laborers understood the systemic violence of their community's living and working conditions. Forced to provide for a family on a shoestring budget, it was abundantly clear to them that a change had to come. Radicalized women such as Mary Septak continued the fight longer than any others, weaponizing their pots and pans.

The next chapter examines the systemic and symbolic violence of the coal mining complexes of the region, the patch towns or collieries. Assemblages of industrial and domestic elements, coal mine operators organized them to be subservient to the intensive functions and interests of mineral extraction. On the one hand, they included the mine workings including industrial complexes of mineral processing and the infrastructure of administration and management. On the other, they included domestic settings: the homes, churches, and community buildings of their laborers, management, and ownership. The latter are not only subservient to the former, they are productive of a social order that supports industrial goals. In a sense, company towns are like factories without a roof, the dwellings of the miners and laborers lining the coal stripping like the batteries of a machine. It is the capacity of the miners to survive each workday despite the traumatic lack of investment in their welfare, and for their families to negotiate these deficient economic conditions, which forms the surplus value captured by coal operation owners.

4

Migration, Part 1

The Social and Material Landscape of the Company Town

David Lynch's 1986 film *Blue Velvet* begins in Technicolor brilliance on a sunny day in a typical American suburb. Early in the film, however, the protagonist, Jeffrey, finds a disembodied human ear in the grass at the side of the road and the halcyon atmosphere of the film is jarringly upended. The discovery of the ear initiates Jeffrey into a journey into the dark and violent realities that underlie the surface of American domestic life. The disembodied ear, social critic Slavoj Žižek says, is a "fragment of reality" that "continually threatens to blow up the settled frame of reality" (Žižek 2001: 129).

Blue Velvet, like many of Lynch's films, is about peering deeply into the violence inside everyday life, revealing the hidden layers that support its operations. I thought of this scene upon reading a newspaper reporter's account of his visit to the site of the Lattimer Massacre the day after the tragic event:

> The shooting occurred at the bend of a dusty road leading from Hazleton and bordered by a rank growth of bush. For a background, however, the affray had a row of half a dozen frame cottages, mean enough in appearance, yet in the little yard in front of each a few bright-hued flowers grow. To-day only a step from these desert blossoms lay a portion of a man's brains, and a little beyond a horrible bundle of gory rags, upon which the blood was still wet. (*Philadelphia Inquirer*, September 12, 1897)

Suggestively juxtaposing the domestic materiality of the company town with the products of horrific violence, the author uncannily reveals the darkness lurking below the surface of the industrial landscape. The reporter exposes residents' quotidian efforts to brighten the landscapes of their living spaces (with "little yards" and "bright-hued flowers") as hopeless efforts at normalcy, merely the "background" for brains and gory rags. It is almost as if these macabre objects grow from the cursed garden soil.

The imagery implies that violence and domesticity commingle in the coal mining company towns of Northeast Pennsylvania. Indeed, these landscapes are suffused with the symbolic and systemic violence of which the Massacre was merely a punctuation mark. As the last chapter demonstrates, the event was a catalyst for confrontations in the villages of Lattimer Nos. 1 and 2. But viewed through the material history of the towns' landscapes, these confrontations also were amplifications of more infrastructural divisions in the town's racial, class, work, and social life. Entangled in transnational racial discourse, industrial imperatives, isolated geography, and sovereign encroachment, its historical development is a rich social context for the aggravation of intersectional class and racial tensions.

Three days after the Massacre, anarchist Emma Goldman gave an insightful speech in Boston, asking the crowd a crucial question (Falk 2003: 286, Shackel and Roller 2012: 769). She theorized:

> If those strikers had been Americans the sheriff would not have dared to fire upon them. But they were foreigners, and foreigners do not amount to anything. The foreigner is good enough to build your elegant houses and your roads, sew your clothes, and do everything for your comfort, but he is not good enough to enjoy the advantages that belong to the heads of the government.

As Goldman suggests, a crucial clue to the origins of the violence comes not from the sequence of the moment, but in the social, political, and material context that allowed the violence to take place. Like Jeffrey's journey in *Blue Velvet*, this investigation demands we consider the domestic landscapes that produced, reinforced, and maintained crucial markers of difference between the citizen and the other.

In her speech, Emma Goldman uses this instance of violence to pose a crucial question: What distinguishes "killable surplus life" from citizenry empowered as "licensed life takers"? During the subsequent trial, the language that justified the Massacre drew from a discourse in which the rule of law, sovereign power, and nation were all on the side of the posse in their containment of a foreign threat. Classified under the rule of *sovereignty*, state power includes the authority to declare life or death, to suspend the rule of law, or to act outside of it. By their presence within territorial borders but outside of citizenship, migrants draw attention to the precarious and vulnerable political condition of all subjects of the state, to the ideological gap between political identity and bare human life (Agamben 1998; Kristeva 1991).

Archaeologists have demonstrated, in a variety of contexts, the power of landscapes to structure social behavior through the imposition of differentiation, discipline, and homogenization. Studies of archaeological landscapes suggest that the powers inherent in social relations are modulated by, and therefore manifested in, built environments, architectural styles, artifacts, and assemblages (De Cunzo and Ernstein 2006; Epperson 2000; Leone 2005, 2010; Nassaney and Abel 2000; Sayers 2003; Shackel 1993;). Architecture and landscape are especially effective as tools of objective violence as they are "relatively permanent, pervasive, and capable of fostering redundant messages" (Nassaney and Abel 2000: 254). In the material culture of racialized difference in the region, a redundantly reinforcing avenue of ideological production is observable in the landscape, architecture, and material culture of the company town.

ARCHAEOLOGY OF A COAL COMPANY TOWN: LATTIMER NOS. 1 AND 2, 1880 AND 1917

> Capitalism builds a physical landscape appropriate to its own condition at a particular moment in time, only to have to destroy it, usually in the course of a crisis, at a subsequent point in time. (Harvey 2001: 247)

In mining districts, "Power and impotence are everywhere juxtaposed, for mining district landscapes are, above all, landscapes of environmental and social control" (Francaviglia 1991: 99). Relations of power, domination, and resistance pervade these social and material environments. The extractive process of coal mining involved a constant struggle to dominate natural forces, materialized in the form of floods, fires, explosions, dust, fetid water, and deforestation. For many observers from the turn of the last century, there was a causal link between the social and the environmental struggles in the region:

> The huge culm and rock heaps, polluted streams, bare and barren hills, cave-ins and strippings, make up the landscape which greets the eyes of these thousands, and if they are polluted in mind and body we need not be surprised. . . . In these regions hardly a spot can be found in the villages and towns that is not cursed. (Roberts 1904: 9)

For those in pursuit of profit from the unwieldy industry, the welfare of ecology, humans, and animals was subservient to the productivity and profitability of the operations.

Company towns, or patch towns, are the foundational landscape features of the anthracite coal industry of Northeast Pennsylvania. They exemplify the interweaving of the social and economic orders that characterize the mining industry. The earliest towns of the region were likely the model for similar arrangements constructed all over the country in the late nineteenth and twentieth centuries (Metheny 2007; Wallace 1978). The Eastern Middle Coal Field, with its extensive feudal power structure, retained the company town system in its most fully developed form well into the twentieth century.

Company towns may seem like a *monadic* form: perpetually totalizing, dominating, and static. However, viewed in the *longue durée*, they reveal a shifting dynamic of violence, contradiction, and movement throughout their history. Capitalist engineering of space serves not only to concentrate production to minimize costs, it also serves the productive function of differentiating and reproducing a striated labor force, physically (that is, systemically) and symbolically, in a manner conducive to productivity (Harvey 1989: 111). An examination of detailed company maps of the town and industrial operations produced between 1878 and the 1940s reveals a seriation to its innumerable transformations. In time, the coal company towns of the anthracite region came to reflect the racialized labor-class hierarchy that prevailed throughout the late nineteenth and early twentieth century.

History of Lattimer Nos. 1 and 2

Calvin Pardee, son of Ario, developed the Lattimer property in about 1866 (ICMAR 1880: 243). The first report of the Inspectors of Coal Mines of the Anthracite Region (ICMAR), published in 1870, reports that "Latimore" shipped a modest 98,723 tons of coal from its single mine slope and breaker, employing 155 men and 32 boys. No fatal accidents were reported, but one serious injury was sustained when James Costule, 35 and married, was injured badly on his back and cut on his head when the car he was riding met another traveling in the opposite direction. By 1872 the ICMAR reported that the colliery employed 118 men inside the mines and 65 as laborers in other outside operations. It also listed the first fatality. On August 19, 1872, Jenkin Rosser, age 27, married with two children, was killed by a premature blast in Slope No. 2. In the remarks section of the report it stated that he lived four days before passing.

The earliest depiction of the company town of Lattimer is the 1873 Beers Atlas (Map 4.1). In this map, the towns appear among a belt of satellite mining communities orbiting the already substantial metropolis of Hazleton. A road heading northward from the city enters Lattimer No. 1 from the west. Company double houses line the town's main street. Evenly spaced, they exhibit the classical homogeneity of company town landscapes. Two larger structures representing the homes of management are visible just to the south of these workers' homes. Topographically, they are situated at a higher elevation, and back from the noise and dust of the mining operations. This location provides them with a panoptic view of the company houses and the mining operations. Clearly marked on this map is the schoolhouse, which played a prominent role in the setting of the Massacre. A rail line and carriage road snake across the former river valley where the coal mining operations are located. Several auxiliary industrial buildings are also present, nucleated around the south side of the mining operations, labeled "Store," "Breaker No. 1," and "Steam Saw Mill." "Breaker No. 2" and an associated mine slope appear to the north side of the operations accompanied by another smaller grouping of houses numbering nine in total, the beginnings of Lattimer No. 2.

In the latter years of the nineteenth century, the landscape of Lattimer Nos. 1 and 2 changed drastically to reflect the new split labor

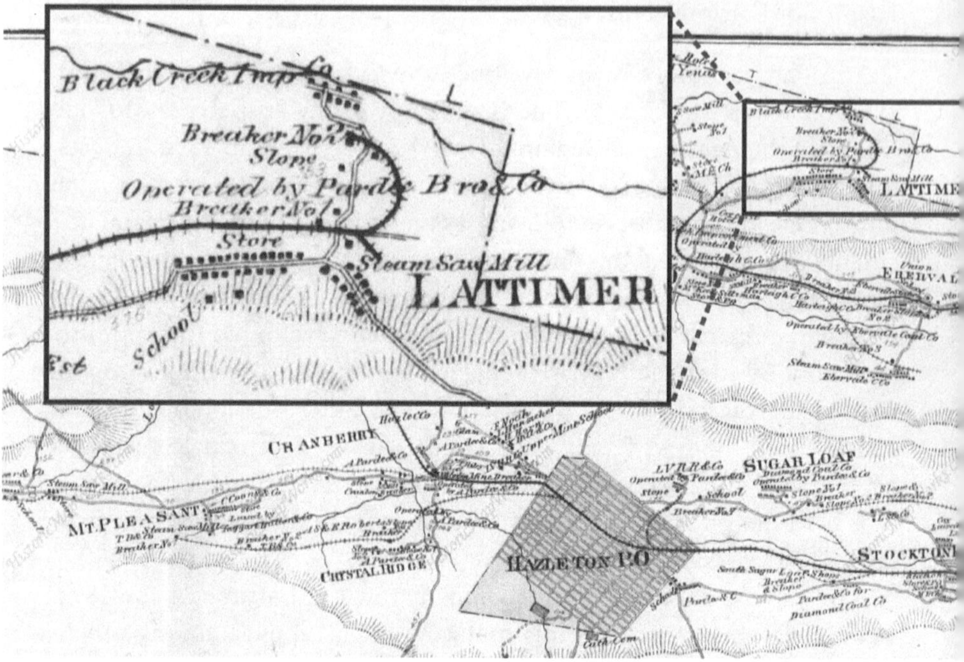

Map 4.1. Detail from the 1873 Beers Atlas showing the early landscape of the Lattimer Coal Company towns (Beers Atlas 1873, adapted by author).

regimes of skilled and unskilled labor coordinated with mechanized production (Map 4.2). Alphanumeric labels on Map 4.2 identify discrete areas discussed in the following section.

A) Lattimer Massacre Site

The site of the Massacre is near the entrance to the town of Lattimer where the road leading from Hazleton splits into Lattimer Main Street to the left and Quality Road to the right (Map 4.2, Area A). Today, in the patch of ground between the split, sits the Monument Stone to the Lattimer Massacre, erected in 1972 by a coalition of organized labor.

B) Company Double Houses, Lattimer Nos. 1 & 2

A total of thirty-two double houses lined the main street in Lattimer No.1. Twenty-two houses were along the main stretch of the road running roughly parallel with the mine operations (Map 4.2, Area B1). An additional three double houses were located along Quality Road or Back Street (B2). Ten double houses lined the road after it took a

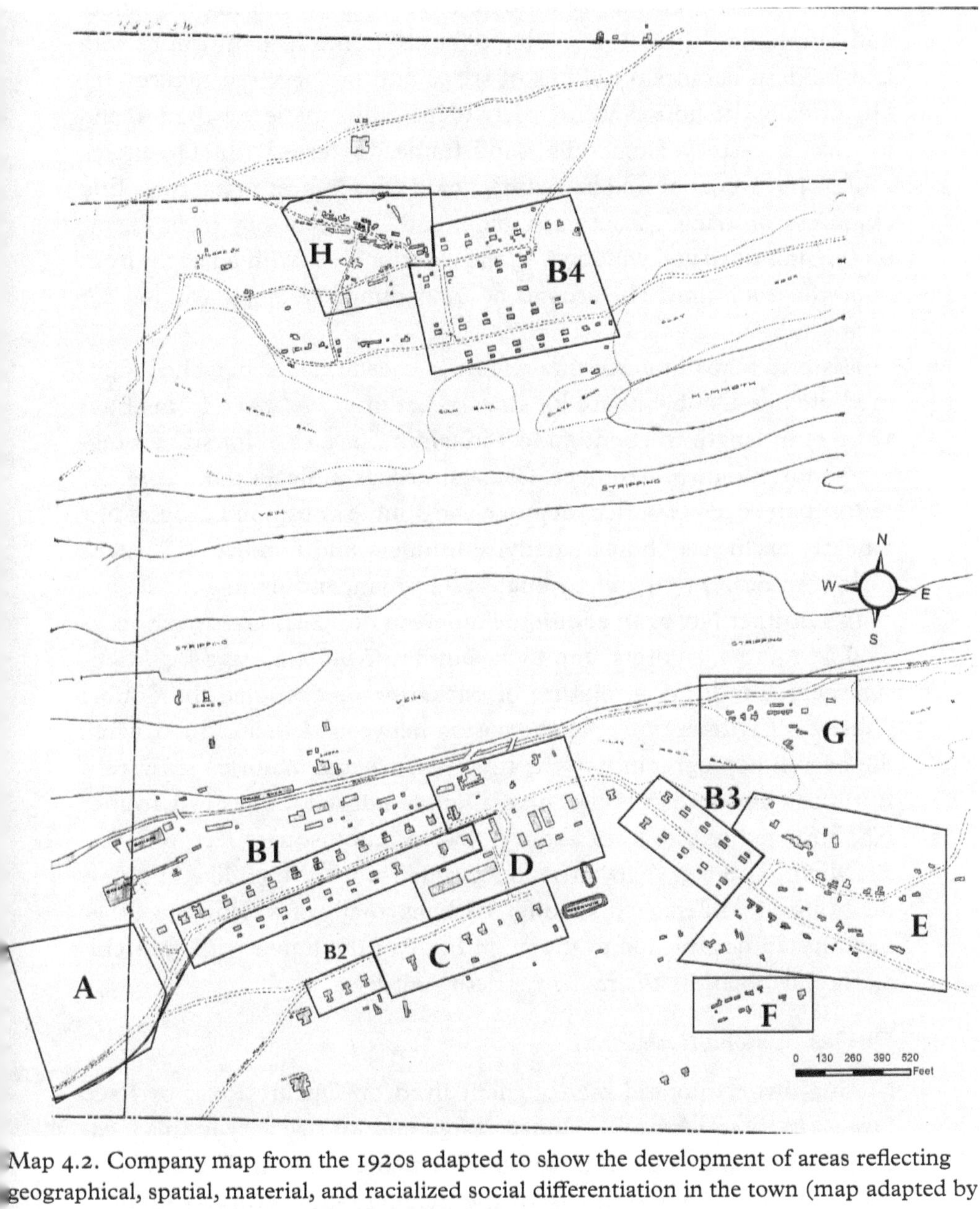

Map 4.2. Company map from the 1920s adapted to show the development of areas reflecting geographical, spatial, material, and racialized social differentiation in the town (map adapted by author, base map from the Lattimer Coal Company Records).

sharp right turn past a high culm bank (Map 4.2, Area B3). Today, private owners have customized the houses, often purchasing both sides and combining them into a single-family home in conformity with late-modern suburban notions of space and privacy (see Figure 1.1). Historically the houses were nearly identical in shape, aesthetics, and lot size: two-story, single-pile wood frame duplexes with side-gabled roofs. They were covered in weatherboard with a sheet metal roof. End chimneys protruded from each gable end. The four-bay front façade on the first floor had windows on the exterior sides with adjacent front doors for each unit. The second floors contained a single window for each unit.

Historic maps indicate that the houses each had a detached summer kitchen. Double house lot sizes in Lattimer No. 1 ran from about 185 feet in length to about 50 feet in width, for a total lot size averaging about 10,000 square feet. Census records indicate that, from the 1880s to the early twentieth century, the double houses in Lattimer No. 1 nearly exclusively housed nativized miners and families of Western European descent, including English, German, and Irish.

In Lattimer No. 2, an additional nineteen double houses were occupied by miners, laborers, and their families (Map 4.2, Area B4). Census records indicate a mixture of backgrounds occupied the double houses in Lattimer No. 2 for the period between 1880 and 1910, when the overall demographics of the town shifted to a majority of Eastern European and Italian households. Double house lot sizes in Lattimer No. 2 ranged in size from about 6,500 to 9,000 square feet, on average slightly smaller than those in Lattimer No. 1. Much like in No. 1, the houses had detached summer kitchens, many of which were either demolished or attached to the main body of the house with an architectural hyphen[1] by the mid-twentieth century.

C) Quality Road/Back Street

Mining ownership and management lived on Quality Road or Back Street (Map 4.2, Area C). House design and lot size vary in this location. This display of architectural ornamentation and diversity contrasted with the homogeneity of the company-built workers' homes, a notable exception to the prevailing material logic of the company town. Between three and four large houses of heterogeneous design populated this street including the large lot of the mine superintendent. Additionally, the company constructed a Methodist Episcopalian church, a schoolhouse, and three double houses. The residences are of Victorian design. The Superintendent's house, located central to the street, is a five-bay, two-story Italianate structure with a low side-gabled roof, and an extensive rear addition features a conspicuous use of Victorian-era trim. The yard measures about 90,000 square feet, or just under 2 acres.

D) Auxiliary Company and Industrial Buildings

Lattimer No. 1 was also the site of a number of auxiliary buildings supporting the town and colliery as well as providing commercial services (Map 4.2, Area D). Company buildings including the company store, storeroom, and office were centralized here. A complex of mule barns and horse stables were located across from the store, relatively close to the mining operations. The Lattimer Supply Company, connected to the Pardee family of businesses, operated two large warehouses across from the company store. These structures housed industrial retail for mining and other industries into much of the twentieth century. A number of light industrial buildings were tucked behind the homes of Main Street adjacent to the mining operations. At various moments in the history of the colliery, they included a foundry, carpenter shop, blacksmith shop, machinist shop, fabricator, and a lumber yard.

E) East Lattimer

A heterogeneous assortment of houses lined Main Street beyond the company double houses (Map 4.2, Area E). Constructed piecemeal starting in the 1870s, these houses reflected company and employee efforts to meet the housing needs of an expanding labor force. Census records indicate that sometime after the 1880s a mixture of Western and Eastern European immigrants began to move into the area, as well as some Italians. For the remainder of the twentieth century the area remained diverse. This neighborhood roughly replicated the population density of the company houses, with lot sizes relatively equal to that offered by the company homes. Evenly spaced and fronting extensions of the main street, these houses echoed the spatial organization of the company housing.

F) "Italian Dwellings," or "the Hill," Lattimer No. 1

In contrast to the spatial arrangements and lot sizes of the rest of Lattimer, two discreet areas in Lattimer No. 1 reflected a different spatial order. (Map 4.2, Area F) A Lattimer Coal company map from 1913 describes this area as "Italian Dwellings." Observable are densely clustered buildings lining a small road or alleyway on the top of a hill south of Main Street. Subsequent map depictions of the area show an organic architecture of house additions, reflecting house expansions and subdivisions to accommodate additional families, renters, or boarders. Today, the landscape and architecture of this area is largely obliterated by selective demolition and recent construction or alteration of existing structures.

G) Canal Street, Lattimer

The shanty settlement of Canal Street occupies a plot of land concealed from Main Street behind a large culm bank (Map 4.2, Area G). The approximately 1.8-acre area occupies a sloping wedge of land between the large bank and a canal that bordered the north edge of the settlement of Lattimer No. 1, which separated it from the colliery operations. Historically, the canal drained water and effluvia from the mines, industrial workings, and domestic occupations. Canal Street terminated at the canal where a bridge carried miners and laborers across the canal to mine workings. In the past, the road turned sharply westward leading into the small settlement. Historic maps from between 1878 and 1941 depict between nine and fifteen houses of varying sizes in the enclave. Historic maps, aerial photography, ground features, and archaeology indicated the presence of additional auxiliary structures on the lot, including privies, icehouses, storage shanties, and coal sheds and, potentially, unrecorded domiciles. According to local informants, the majority of structures were demolished by the late 1960s. Presently, two heavily altered historic structures as well as a recently constructed (post-1970s) house stand in the area. Older residents of the town remember little to nothing about the occupants, structures, or landscape that made up the settlement. Most new residents we spoke with did not know that it existed.

H) Shanty Enclave, "the Village," Lattimer No. 2/Pardeesville

In the second half of the 1870s, a number of ephemeral and semipermanent structures were built in the northwest corner of town. A map from 1885 shows the St. Nazarius Catholic church (Map 4.3), constructed a year before, surrounded by a heterogeneous assortment of structures organically clustered around it. Structures included several long barracks-like buildings accompanied by smaller constructions. These included both dwellings and structures serving auxiliary domestic functions such as kitchens, privies, and storehouses.

In the decades that followed, population density in the enclave increased, with houses built to fill much of the available space. Company maps from the second decade of the twentieth century depict as many as forty-two individual structures concentrated around the intersection of two alleyways dominated by the church (Map 4.4). Lower density housing spread throughout the west side of Lattimer No. 2, scattered among fenced-in empty lots containing gardens and animal pens. A lifelong resident of the town described the formation of the shanty enclave on the west side of Lattimer No. 2:

> You see where the coal company houses stop? This is where the dividing line is. And the reason for that is prior to the Italian nationality moving in to this area, even in Lattimer Mines, what we had was the coal company houses. They were built by the coal company . . . the English were the first ones to move over here, in those homes. . . . So if you look to the west. . . . from these coal company houses, what do you see? All these irregularities. And the reason for that is when the Italian people came here they acted as squatters. There was no system from the coal company to control anything. And so they moved in there, and I was told by some of the old people that they cleared the land, clear-cut the brush and just began tilling the soil. . . . the irregularity. . . . it's not like with the coal company houses. . . . They just went in there and they just did basically whatever they wanted to do. There was no control from the coal company. (MD, interview, March 21, 2013)

The resident notes an invisible line dividing the two districts of the company town by space and architecture; he also suggests that this divide is more than simply material but also social. In this interpretation,

Map 4.3. Detail of 1885 map depicting the St. Nazarius Catholic church and other buildings in Lattimer No. 2 (map adapted by author, base map from the Lattimer Coal Company Records).

Map 4.4. Detail from a company map dating to the second decade of the twentieth century showing the Lattimer No. 2 shanty enclave superimposed over present satellite image of Church Street, Pardeesville. The arrow indicates the study area (map from the Lattimer Coal Company Records, aerial photo from Google Earth, October 12, 2014; overlay by author).

peripheralization also brought freedom when he notes that "they just did basically whatever they wanted to do. There was no control from the coal company."

At other moments, observers have addressed this irregularity in different terms. In 1898 a magazine reporter and artist for the *Century* magazine visited Lattimer to report on the "foreign element" involved in the Lattimer Massacre to its largely white middle-class readership (Hambidge 1898; Rood 1898). Their agendas are transparent in the reportage, guiding readership to draw racialized character traits from shanty landscapes. Their arguments for immigration restriction are embedded in these descriptions, warning that the Slavic and Italian immigrants are "much more dangerous to the body politic than the excluded Chinese; for not only are they eager to work for wages on which an English-speaking family would starve, but they are superstitious and murderous" (Rood 1898: 811). The artist Jay Hambidge is particularly receptive to aesthetics, explicitly articulating the epistemological connection between materialities and "mentalities" (1898: 824–825):

> This is a place to be described by metes and bounds and degrees of instruments, and to do it would tax the ingenuity of the best of surveyors. Can you read character from handiwork? If so, this would be a place to practice your art. Each little house, with the boxes, cubby-holes, and fences about it, has been built by the man who lives in it. And he is a laborer, a struggler for mere existence, not deft in the use of tools, nor with an eye for the symmetrical, nor with an appreciation for anything beyond the most primal facts of living.

Hambidge describes the settlement as "a hazy-looking mass with many poles sticking out of it" (1898: 824) consisting of:

> *the queerest structures*, some of them not much larger than dog-kennels. . . . *There is no sewage system here*, and the alley is the dumping-ground for all offal. At every few steps of this winding, reeking way are little openings leading into other' passageways, not much wider than will permit a man to walk through.

And elsewhere Hambidge (1898: 825) relates:

> The roofs of the buildings slant at all angles, with no two sides of the same length or deflection. One portion will have eaves,

> while its companion will scorn the luxury. The same incongruity prevails everywhere. Some of the small openings used for windows are high, while others are low. One door will open in, and another out. The hinges have evidently come from the company scrap-pile, and the staples and latches and locks from the same source. Some of the roofs have shingles, others weather-boards, while others are formed of great pieces of rusty sheet-iron.

Ganaro Volco, an Italian miner and shoemaker sat for one of the portraits drawn by Hambidge that accompanied the article. Although they could not exchange ideas in English, Hambidge (1898: 828) reads into the terms of his fellow craftsman the confirmation of his opening thesis:

> A shoe, he said, when made by hand,—and though he couldn't express himself clearly in English, I understood,—*reflected the character of the maker, and, according to the strength or weakness of character, was good or bad*, and in such measure was art. The same article made by machinery—a shrug of his small shoulders [italics added].

Hambidge appropriates the voice of the shoemaker to authenticate his determinist materialism: that objects reflect the character of their makers.

Hambidge and Rood perceive a built environment attributable only to the immigrant laborers. Their understanding of the landscape is isolated from the many historical factors that contribute to its character. The authors are entirely ignorant of the structural role the mine owners have taken in engineering its existence and maintaining its conditions to their advantage through regimented racial hierarchy, poverty, and debt. Moreover, they cannot see the dynamic character of the enclave, which has already been in a state of constant improvement and adaptation by its long-term residents for over twenty years. For these authors, this is a place with no history . . . and no future they can imagine.

The 1898 articles by Hambidge and Rood played a significant role in the broader narrative history of the Lattimer Massacre. During the Austro-Hungarian government's claim for indemnity for the victims of the Massacre, the U.S. government assigned Assistant Attorney General Henry Hoyt to report on the case for the government. Hoyt's report includes the articles as evidence for the government's case against distributing indemnities. Ignoring the populist persuasion rhetoric of these articles, they are provided as authoritative evidence: as a "recent statement of [the immigrant's] manner of living, customs and characteristics." Hoyt, based upon the 1898 article, finds that the strikers of foreign birth "are ignorant, and from our point of view uncivilized" (Hoyt 1898, quoted in Palmer 1913: 193). He continues:

> Compared with such people of our own, they represent a status of civilization to find which generally we should have to go back several centuries. Their manner of life, their standard of living, their domestic relations and moral sanctions are all crude and defective in comparison with any proper standards.

Hoyt relents, however, in that he sees the possibility for the future "amalgamation" of the foreigners in lieu of their character; in their thrift, their desire to become citizens and own homes, their affection toward children, their attachment to church, and their "loyal[ty] to those who help them or show them kindness" (Hoyt 1898, quoted in Palmer 1913: 193).

Today the landscape of the shantytown is still visible if you look closely (for example, see Map 4.4). The unique spatial arrangement of houses has been heavily obscured by the actions of residents, real estate companies, and government, rendering it nearly invisible to the casual observer. As discussed in Chapter 7, a close examination of historic maps and archaeology reveals that selective abandonment, gentrification, and systematic renewal altered the landscape of the town to reinvest it with values amenable to late capitalism. Archaeological remains, however, attest to this history.

Archaeology of Church Street, Lattimer No. 2 (Map 4.2, Area H)

The landscape that once gathered around the St. Nazarius Catholic church in Lattimer No. 2 retains the general outline of its historic setting, albeit with significantly altered aesthetics and density. Town occupants demolished the church in the 1950s, building a newer church at the top of the hill overlooking the town. In the process, they discovered that it was constructed of dynamite boxes and scrap collected from the colliery junkyard.

Today, the arrangement and aesthetics of contemporary houses in the district do not look different from the typical suburban landscapes constructed throughout the country during the latter half of the twentieth century. However, beneath the surface treatment of vinyl or aluminum siding applied to some of the standing structures, shanty architecture dating to the early part of the century is largely intact in some structures. This intimate landscape is also visible in the spatial relationships between some houses, with additions that interlock at oblique angles to maximize the use of space. Moreover, in many areas the archaeological remains of houses are intact in subsurface deposits.

In the summer of 2013 the Department of Anthropology ran a field school in two lots along Church Street in Lattimer No. 2, since renamed Pardeesville (Map 4.4, red outline). The study area covered a total of 0.13 acres, or about 5,200 square feet, and is bordered by an alleyway to its west that historically separated the study area and a narrow alley. To the west of this alley was the site of the church. In fact, this is the alleyway ascended by Jay Hambidge to enter the town in the 1898 article (824):

> We are soon at the opening of a lane much narrower and more crooked than the one we have just traversed. It leads into the patch, and up this lane we drive. Then up another lane, more tangled than the last; and this is the principal street of a settlement of the queerest structures.

A single subdivided domicile, now demolished, stretched across the north end of three lots fronting Church Street. Company and real estate maps from between 1885 and the late 1940s capture the rough outline of this evolving structure. Each lot had its own small backyard. At the time of the survey, the lots were open grassy areas with some pockets of dense brush and small trees. No features were visible on the surface of the ground other than a few slight depressions.

Field school students conducted a shovel test survey of both lots. Twenty-three shovel tests were plotted at 10-foot increments down the center of each lot. We recovered a total of 764 artifacts in the shovel test survey. The shovel test survey of the lot to the east, owned by W. Klatch in 1941, followed by the DeLorenzo family for a period of years, yielded an unusual signature visible in the apparent topography of backyard soils. Shovel test soil profiles in the rear one-third of the yard revealed severely depleted yard soils. A conversation with a former resident of the house who has lived in Pardeesville for eighty-six years explained.

When the family moved to a new residence on the newly constructed Upper Street in Pardeesville in the 1950s, they mechanically scraped and transplanted the valuable rich garden soils. To this day she still maintains a garden with Italian vegetables such as tomatoes and basil in these same yard soils.

Given the soil depletion identified in the DeLorenzo plot, we excavated all thirteen test units in the western lot. I identified the lot on a real estate map from the 1940s as occupied by "D. Simone" and family. Michael and Eleanor Diblasi purchased the house from the Hazle Realty Company in 1959. A total of twenty-four features were identified in test units including foundation walls, utility pipes, concrete pads, postholes, and a privy. Figure 4.1 illustrates the locations of selected test units and features identified during the survey. Selected features are discussed below and in appropriate sections in Chapters 5 and 6.

In contrast to the enforcement of rigid homogeneity of company double houses, shanty occupants were free to alter the site plan of houses in response to changing needs or sensibilities. Hambidge describes one such house, which had been expanded with so many additions over the years that it resembled "a great black worm" (Hambidge 1898: 827). This organic additive architecture is visible in the archaeological record. Architectural features identified in Units 9, 10, 11, 12, and 13, viewed in combination with company and real estate maps from between 1885 and 1941, depict a radically organic vernacular architecture on Church Street.

Feature 1 is a 5 foot × 5 foot privy located in the southwestern corner of the rear yard of the building. The privy was wood-lined with vertical slats broken off about 3 feet below the surface. Support posts of varying widths (0.4 ft. to 7 ft.) were intact at the corners starting at a depth of 1.5 feet below the surface. From the contents of the privy, the date of construction could not be ascertained as it had been cleaned and refilled at least once in its use-life. From an analysis of the contents, it appeared residents used it as a receptacle for a single large clearance from the house at the transfer from the Simones' tenancy to the Diblasis,' sometime around 1959. The contents of the privy are discussed in detail in Chapter 5 within an appropriate context of mass consumption and the rise of late modernity.

The obsolescence of the privy likely corresponds to a variety of infrastructural improvements uncovered at the Church Street site. Features

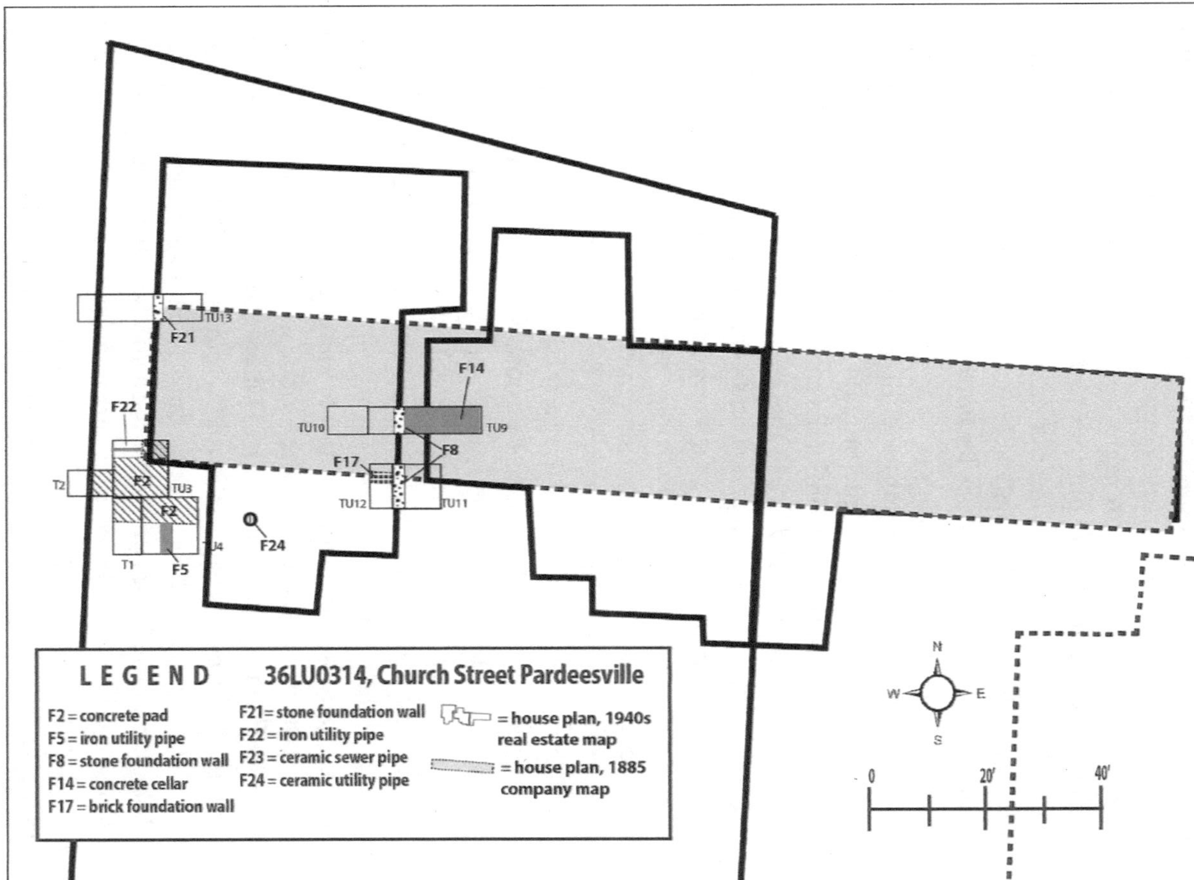

Figure 4.1. Architectural features (a brick foundation and a stone wall) and selected features (Feature 17 and Feature 8) identified during the 2013 archaeological excavation at Church Street, Pardeesville, recorded during the 2012 survey and superimposed over house plans from 1885 and 1941 (map adapted by author, base map from the Lattimer Coal Company Records).

5, 22, 23, and 24 all represent house improvements directed at draining wastewater and effluvia away from the house. Features 5 and 22 are cast iron pipes leading from the house. Feature 22 ran westward out of a rear addition, into the alleyway. Feature 5 led toward the back of the house, perhaps to a cesspool. Diagnostic artifacts found in feature soils, including pipe trenches, suggest that these improvements were installed by a succession of owners of the house throughout the first quarter of the twentieth century. Feature 2 was a concrete pad poured at the corner of the house to drain wash water. Informants in town describe the daily after-work outdoor baths miners used to remove the dust and grime before entering the house (BO, MG, NG, interview 2012). In addition to these daily ablutions, the concrete pad provided an impermeable sanitary surface for household chores such as washing clothes and other household items.

Superimposing the floor imprints of successive house depictions reveals that the footprint of the original structure remains intact, to some extent, within the footprint of the structure throughout its history (Figure 4.2). The first structure in this area, present by 1885, is a long barracks-like building. It is depicted on an early map as rectangular and without any additions. Feature 17 may represent a remnant of the walls of this structure, encased within more typical shanty architecture of broad stone walls built sometime after 1900. Feature 17, composed of mortared brick with a width of about 0.7 feet, angled slightly northwest in conformity to the direction of the main body of the building depicted on a map surveyed in the 1940s. By 1900, occupants had added two additions to the front, one of which was separated from the main body of the house with a small gap. In the years that followed, between 1900 and 1941, residents made a series of cumulative additions to the front and back of the house.

In the 1940s a literal rupture in this pattern of cumulative growth separated the two households in the form of a narrow (≈3 feet) alleyway (visible in Figures 4.1 and 4.2). The DeLorenzo family, who occupied the house to the east briefly in the 1940s, recalls the narrow passage between the houses. It is theorized that the original 1885 structure was demolished or burned sometime after 1900 and two structures rebuilt on its footprint. This time, however, a narrow alleyway was left for privacy and differentiation between the westernmost unit. Another explanation is that the two structures were purposely separated to differentiate

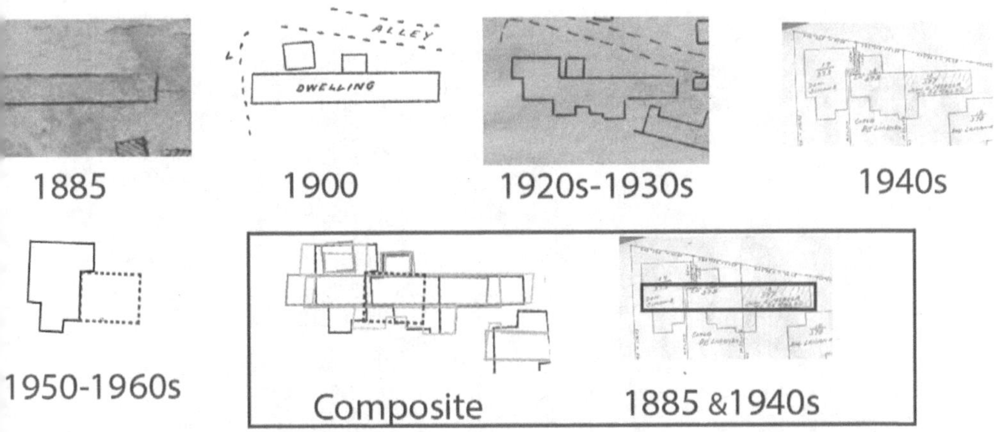

Figure 4.2. Structure floor plans from between 1885 and the 1940s (figure by author).

the units. Test units investigating the alleyway (TUs' 9, 10, 11, and 12) failed to turn up any subsurface indications of the passage. A 2-foot-thick stone foundation wall (Feature 8) was uncovered along the eastern wall of the Simone/Diblasi residence. As mentioned above, to the west of Feature 8, remnants of the 1885 brick foundation in the form of Feature 17 were identified, interrupted at less than a perpendicular angle where it was cut off by the construction of the stone wall (Figure 4.3). The angle of this masonry conforms to the outline of the 1885 building. Where the alleyway should have been, directly to its east, we identified a concrete cellar (Feature 14) packed with architectural and domestic debris ranging in date from the turn of the century to the 1970s. As it dates to a later period, this feature is discussed in further detail in Chapter 6.

PARSING THE MONAD OF THE COMPANY TOWN: HETEROTOPIAS AND "THE CAMP"

Characterizations of company town landscapes in academic analysis and popular culture present them as among the purest form of rationalized control imposed upon social, spatial, and economic material reality (Dinius and Vergara 2011; Francaviglia 1991; Green 2010; Mulrooney 1989). In the manner in which they direct all social life toward a rational goal, uprooting, harnessing, or replacing traditional organizing

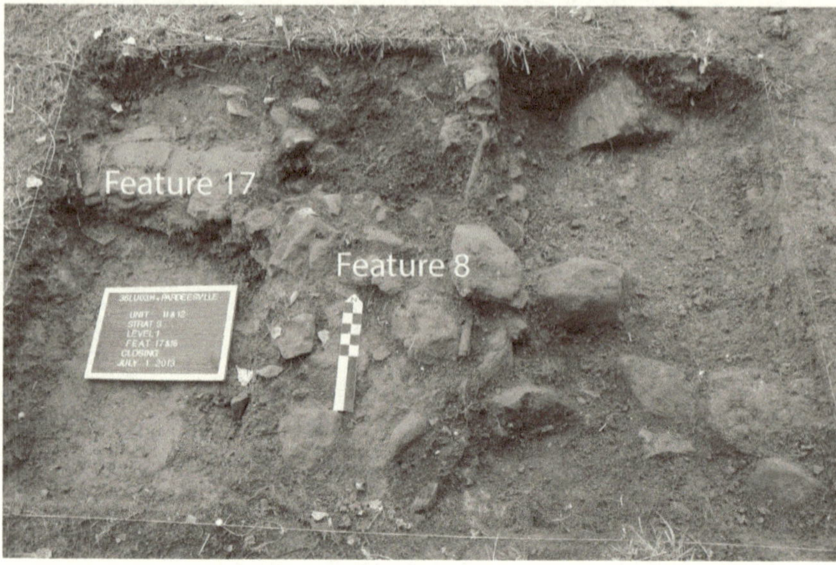

Figure 4.3. Feature 17 and Feature 8, remnants of the 1885 brick foundation and stone wall, respectively (photo by author).

units with rationalized financial goals, company towns exemplify the rupturing of the past and present central to the historical developments of modernity and capitalism. Endowed with a belief in the inherent plasticity of social order, company towns exemplify the political project of modernity, the possibility for engineering social life toward ideological goals (Herod 2011). For capital, producing a landscape such as the company town is a fixed investment, a system of networked machines to aid in concentrating production to best capture surplus accumulation and labor power (Harvey 1985, 1989; Marx 1977: 492). It also serves the social function of differentiating and reproducing a labor force in a manner conducive to productive surplus (Harvey 1989: 111).

As these formations play out in material history, however, contradictions abound. During the rise of industrial capitalism, the conflicting strains of national identity and a voracious demand for unskilled labor placed migrants in the paradoxical position of being both appropriated and restricted. Given the contradictions of company town life, a monadic characterization may seem overdetermining. The many separate districts of Lattimer Nos. 1 and 2 present heterogeneous positions and conditions of exception that challenge a simplified notion of rationalized order. For example, the homogenous ordering of the company houses is opposed by *both* the exceptional heterogeneous spaces of the management enclave *and by* the isolated peripheral shanty settlements. The low-density settlements that begin just at the borders of the company double houses seem to echo their density and arrangement, while also reflecting the organic architecture and growth of the other more concentrated ethnic enclaves. The spatial arrangement of the three shanty settlements, moreover, mirrors the spatial organization of the larger town in miniature, with houses facing inward lining a central thoroughfare. Both, in different scales, materialize a community defined by shared communal space and access. The shanty enclaves, however, in their isolation and the inscrutability of the communal spaces they form, are also invisible to the surveillance of company eyes. This invisibility is complemented by the anarchic or organic quality of their infrastructure, their resistance to the documentary efforts of surveyors, and their operation outside of the corporate infrastructure.

To explain the nature of contradictory spaces, we can turn to theoretical ideas from Foucault (1984), Agamben (1998) and Žižek (2013) that assess the functions of contradiction in regard to ideological power. In the essay "Of Other Spaces," Foucault (1984) suggests that circumscribed spaces suffused with power and social division may represent a counterexample to simplistic and totalized notions of space, described as utopias. These circumscribed spaces he describes as *heterotopias,* which represent a reflection of real power relations in the world. He suggests:

> There are . . . probably in every culture, in every civilization, real places—places that do exist and that are formed in the very founding of society—which are something like counter-sites, a kind of effectively enacted utopia in which real sites, all the other real sites that can be found within the culture, are simultaneously represented, contested, and inverted. . . . The heterotopia is capable of juxtaposing in a single real place several spaces, several sites that are in themselves incompatible. (1984: 24–25)

The many districts of the company town of Lattimer incorporate such a space in which total administration is simultaneously confronted with its contradiction.

How does this function politically? In other words, how do the excluded affect the totality? Agamben imagines the material locus of biopolitics in a way that functions as a fitting counterpart to the explanatory scale of the heterotopia. In *Homo Sacer* (1998), Agamben suggests that the concentration camp, the detention facility, and the refugee camp represent primal spatial materializations, described by him as the *nomos* or the *paradigmatic space* of political modernity in the twentieth century. According to Agamben, the spatiality of the camp at first operated as a panicked (and yet carefully engineered) response to the challenge of diversity brought on by mass immigration. However, the exceptional space of the camp also served to consolidate the national body through the intensification of governance simply by its presence just on the outside of prevailing logic. As Agamben defines it:

> The camp is a piece of land placed outside the normal juridical order, but it is nevertheless not simply an external space. What is excluded in the camp is, according to the etymological sense of the term "exception." . . . *taken outside*, included through its own exclusion. But what is first of all taken into the juridical order is the state of exception itself. . . . The camp is a hybrid of law and fact in which the two terms have become indistinguishable. (Agamben 1998: 169–170)

To the constellation of refugee camp, detention facility, and concentration camp, we can add the shanty town, a space conscripted for migrant laborers on the outside, but also central to the tightly controlled boundaries of the engineered space of the company town. Agamben's consideration has deep political implications today. The paradox of his proposal is that exceptional spaces serve the key function of defining *within the order* what is distinguished as inside from outside, thereby establishing the authority of sovereignty. In turn, the formative presence of exception at the top and the bottom of hierarchy, in turn, "inaugurates a new juridico-political paradigm in which the norm becomes indistinguishable from the exception" (Agamben 1998: 169–170).

THE MIGRANT AND THE MACHINE

In the years leading up to the Lattimer Massacre, the mine operators and the middle class of Hazleton benefited from stable profits coming out of an industrial-scale coal extraction that was dependent upon mechanization and labor exploitation. This privilege afforded them comfortable, stable domestic lives, isolated from the precarity and danger of the company towns. Threatened by unsteady markets and the increasing organization of longer established craft labor, the industry turned to a combination of mechanized production and unskilled recent immigrant labor. For a time, this underclass of surplus labor was far less demanding of wages, stable income, and safety measures than earlier generations of nativized labor. Systems of objective violence differentiated these populations through domestic landscapes and racialized characterization to justify this split labor force.

Much like other sectors of industrial capitalism in the late nineteenth century, coal operators integrated mechanized processes and unskilled immigrant labor into coal extraction to maintain escalating profits (Aldrich 1997; Aurand 1985; Roberts 1901). As it functioned in other industries, unemployment, de-skilling, and mechanization also served to destabilize established labor regimes (Paynter 1988). Two new laboring processes integrated unskilled labor with mechanization to capture surplus production: stripping, and washery coal. Washery operations used water and mechanization to recover coal mixed with other discarded by-products from the many waste piles ubiquitous in the region. For miners traditionally paid only for the usable coal extracted from cars, the exploitation of discarded washery coal by operators was a kind of theft, hence one of the mottos used during the 1900 strike exhorting miners not to "handle washery coal" (Roberts 1901: 212–213).

Stripping involves the mechanized extraction of coal seams by shoveling surface soils. Depending on the thickness of overburden, stripped coal could be produced by companies at half the cost of shaft mining (Roberts 1901: 22). Available by the 1870s, steam shovels facilitated the work of stripping in place of the handiwork of traditional shaft mining. In essence, strip mining bypassed the delicate constraints of craft skill entailed in the shaft mining process. Likewise, it destabilized the workforce by operating with a shifting workforce. This labor was worked by day laborers under a foreman, and largely undertaken by new immigrants such as those that struck during the Massacre. In 1904, a commentator reported that Italians have "nearly monopolized the work in stripping mining" (Roberts 1901: 33).

Calvin Pardee, owner of Lattimer Mines, introduced strip mining to the anthracite region in 1874, and by 1881 still operated the most extensive stripping operations (ICMAR 1881: 279). By 1890, and likely earlier, stripping was introduced extensively at Lattimer, along with washery operations (ICMAR 1891). As a result, this unskilled labor force peripheralized at the edges of town grew increasingly central to operations. Worker demographics reveal the moment of highest growth in the shanty enclave coinciding with the introduction of mechanized practices into the operations of the Lattimer Colliery. Moreover, the rise of labor radicalism among older residents in the 1880s coincides with this decrease in skilled labor (underground mining) and a steep increase in other forms of unskilled and ephemeral labor. This system reached its peak in 1892, when the employment of skilled miners was at an all-time low in proportion to other forms of colliery labor. Consequently, a temporary precipitous drop in stripping and washery labor between 1897 and 1899 may reflect the radicalizing of the immigrant workforce following the Massacre or the reported exit of migrants from the company in the aftermath of the incident.

The shanty enclaves that formed at the margins of the company town are direct materializations of these industrial innovations. The initial construction in the form of the brick foundations of a barracks-like building may have been built by the company in the 1870s to accommodate the introduction of new immigrant labor. Archaeology reveals that parts of this original structure were integrated, in some form, into the shifting architecture of the site until its demolition in the late 1960s. A magazine article from this period equated the racialized character of the inhabitants with the heterogeneous character of shanty architecture, neglecting to note the role of the company in creating the community, maintaining its differentiated character, and profiting from its surplus production. The discovery of this foundation is a literal challenge to the refusal of Hambidge (1898) and Rood (1898) to recognize the history of the settlement.

CONCLUSION

The shanty enclaves functioned as autonomous and unordered space within the margins of the company town, resisting its order and visibility, although they were essential to the broader system of its productive function: supplying the colliery with a steady supply of surplus labor. Archaeology and archival research suggest that far from being ephemeral, these settlements were constructed fewer than fifteen years after the initial founding of the towns, and remained a permanent, albeit shifting, fixture on the landscape. This is quite different from other scholarship that presents such satellite communities as alternative spaces for those who seek to operate outside of capitalism (Goddard 2002; Hardesty 1998; Matthews 2010: 138–141).

In the *longue durée*, the shanty enclave models for all workers the presiding conditions of precarity and self-sufficiency demanded of them by the capitalism of the late twentieth century. Indeed, labor precarity today is no longer restricted to the proletariat, but is a fact of life for all workers, including immaterial and intellectual laborers (Lazzarato 2012). Slavoj Žižek (2013: 1003) reminds us that the inscription of the excepted is an implicit element of capitalist exploitation. The surplus army of labor workers are "not simply outside the circulation of capital, they are *actively produced as not-working by this circulation*. . . . They are not simply not-working, their not-working is their positive feature" [italics mine] (Žižek 2013: 1003). In essence, contradiction such as that of exclusion is implicit to the functioning of capitalism itself and amplified in its more complex forms (Harvey 2014; Jameson 2011).

The peak of the industry in the region came during the period of the First World War, with market demand high and profit returns from technological investments providing healthy returns. The period during and after the war saw a decline in new immigration to the region owing to interruptions in mobility and, in 1924, the passing of immigration restriction laws. The tapering of immigration to the region slowly altered the workforce. It also pressured the immigrant community to adopt their place within a "hyphenated-American" pluralism. Archaeology revealed resident and community efforts to improve sanitation and private space throughout this period, adapting temporary residences into a permanent community. Excepted from the homogeneity rigidly enforced on company homes, the lots in the shanty enclave evolved organically into individualized family lots. A 3-foot alley appears briefly, cleaving one domicile from another, perhaps reflecting evidence of growing concerns for privacy, property ownership, and differentiation.

Nonetheless, the struggle between labor and capital continued, albeit in altered form. In the next chapter, I introduce the beginnings of an era of softer, more diffuse forms of coercive power and control. These came in the form of a diabolical solution engineered by elites to the twin problems of productive surplus and social instability. This history helps explain the presence of contradictory elements—poverty and material abundance—in the privy deposits excavated from the shanty at Church Street.

5

The Archaeology of Machinic Mass Consumerism and the Logistics of the Factory Floor in Everyday Life

> Cut loose from the traditional communities that circumscribed their functioning, they have begun to wander everywhere in a space which is becoming at once more homogenous and more extensive. Consumers are transformed into immigrants
>
> *Michel de Certeau,* The Practice of Everyday Life *(1984: 40)*

> There is nothing as mysterious as a fact clearly described.
>
> *Garry Winogrand*

THE CONSPICUOUS SYMBOLISM OF THE AMERICAN FLAG, LATTIMER, 1897 AND 1917

The American flag appears twice in my documentation of labor struggle in the Anthracite Coal Region of Northeast Pennsylvania, first as tragedy and second as farce. In juxtaposition, they divulge a historic irony revealing of the drastic changes that occurred in the region and the nation following the Massacre. In its first appearance, it was carried by the striking laborers as they marched to Lattimer in 1897. Unarmed, their only defense was that offered by the flag. At their first confrontation, it was taken from them, torn to pieces, and thrown to the ground. In the court case that followed, the symbolism of the flag was touted as a symbol of country and race, law, order, or the protected rights of citizens. Deputies defended themselves in suggesting that the foreigners had no right to wield the flag, much less gather and protest for their rights.

Twenty years later the flag appeared again, this time in striking contrast to the previous incident. On April 6, 1917, the United States government committed itself to full participation in the First World War. The antiwar movement that flourished in response was inextricably linked with the labor radicalism of the time. In these troubled times, the war was perceived by many observers as an extension of domestic class struggle as the wealthy sent the working class to death in battle for no other reason than to protect the stability of their economic interests (Zinn 2003: 359–379). By May 9, 1917, barely a month after the start of the U.S. involvement in the war, a flag-raising ceremony was planned for Lattimer at the No. 5 colliery, one of many such ceremonies coordinated at coal mines and other industrial centers across the country (Figure 5.1). At this event an officially sanctioned raising of the flag was accompanied by a telling didactic exercise. Local schoolchildren sang the anthem; children who likely exhibited superior command of the national language and rituals their foreign-born parents had not yet gained. Performed in an isolated industrial community, this instance of political theater was part of the larger efforts of the official propaganda campaign of the First World War.

It is not recorded if the event was well attended or received. Similar events, however, were conducted in small towns throughout the region, some accounts of which were reported in industry circulars. An account from the same day conducted at the Butler Colliery in the northern coal field town of Pittston describes one such event:

> Fifteen hundred persons attended the flag-raising exercises at the Butler colliery of the Hillside Coal and Iron Co. near Pittston, Penn., on May 9. Rev. S. Ezra Neikirk, of Pittston, was the speaker of the day, and Rev. Father Zimanski delivered an address in Polish. (*Coal Age*, May 19, 1917: 876)

These two anecdotes, from 1897 and 1917, reveal a striking paradigmatic shift in corporate and political approaches to governing American pluralism. The radically different responses to the image of new immigrants rallying beneath the nation's flag by authorities of capital, state, and law and order present a telling example of the rapidity with

Flag Raising
Exercises
—at—
Lattimer No. 5 Colliery
Lattimer Mines, Pa.

Wednesday, May 9, 1917
at 4:00 P. M.
Held Under the Auspices of
Employees of
Pardee Brothers & Co., Inc.

Figure 5.1. Program from a flag-raising exercise held in Lattimer on May 9, 1917, at the beginning of U.S. entry into the First World War; it was organized as part of a coordinated wartime effort to court the patriotic support of foreign-born workers at sites of strategic wartime importance (photo by author).

which migrant treatment can change in response to historical events, in this case the First World War.

At the same time, coercive measures had not been dispensed with by 1917 but held in reserve or legitimized through the juridical realm. During the propaganda campaigns of both World Wars, coal companies were granted sovereign support in their efforts to regulate workers into new regimes of productivity and discipline beyond the coercive measures of the Espionage and Sedition Acts. A report on the war effort from May 19, 1917, avowed:

> Mine operators at Hazleton Penn have let it known that such of their employees as make a habit indulging in sprees following pay days will be marked as slackers and recommended to the Government as good material for military service. (*Coal Age*, May 19, 1917: 876)

Accelerating production and regulating consumption did more than enhance wartime efficiency and stabilize profit. It also served to synchronize the habits of citizens through the patriotic regimentation of everyday life. In this way, consumption and production were attached to American political ideology by 1917.

In the period after the First World War, capitalists, intellectuals, government administrators, and advertisers proposed the development of intensive forms of consumerism among the masses as a parsimonious but diabolical solution to two major crises confronting the nation, one of a financial nature, the other a matter of unstable governance. The first crisis entailed the need to meet unmet demand for the surplus of commodities flowing out of mechanized industrial production. The second crisis, one of sovereign power, concerned growing anxieties over the nation's capacity to maintain control over its new immigrant populations and radical elements of the labor force. Discourse of the time evokes the belief that the distribution of material abundance, or simply the desire for its consumption, would ameliorate, conceal, or divert attention from the material conditions that lead to struggle by radical or unassimilated elements of the working class.

Writing in this time, Walter Benjamin astutely observed that "the mighty seek to secure their power with blood (police), cunning (fashion) and with magic (pomp)" (Benjamin 1999: 133). This chapter deals with the second element in Benjamin's trichotomy, that of cunning and fashion. For Benjamin, elite control of fashion represented more than simply administering to consumer taste. Rather, it implies mastery of time and history, of the cycles of memory and forgetting accreted in the material world of objects and landscapes (Buck-Morss 1989: 97–99). The fashion cycle embodies the manner in which the temporality of industrial modernity touched the everyday life of producers and consumers exemplified in things such as the accelerating product cycle, the assembly line, and the new media of communication and transportation. Engineered to operate in concert with developments in infrastructure, media, aesthetics, materials science, logistics, work processes, and financial and governmental reorganization, mass consumerism enunciated in the lives of all citizens one quotidian material nexus of a massive, heterogeneous but integrated machinic system of profit capture. In effect, it insinuated a softer, more liquid form of social control, arguably no less potent than the more coercive forms of the past.

PROHIBITING THE ENJOYMENT OF THE OTHER: CONSUMERISM AND DISCIPLINE IN THE ANTHRACITE REGION

In *Fragments of an Anarchist Anthropology*, David Graeber (2004: 74) suggests that, "if one wishes to inspire ethnic hatred, the easiest way to do so is to concentrate on the bizarre, perverse ways in which the other group is assumed to pursue pleasure." He posits here that envy and ressentiment are at the center of the puritanical efforts of capitalism (Graeber 2004: 74; Žižek 1992, 2008). Žižek proposes that human desire is always framed, or rather blocked, by the imagination of how the other is "able to enjoy" the object of pleasure (2008: 87–90). Envy is incited in the puritanical subject not by the possession of a prized object, but rather, "to destroy the Other's ability/capacity to enjoy the object" (Žižek 2008: 90). Moreover, this ressentiment likely plays no small part in the historic movement of capitalism to rid society of "collective, festive consumption to the promulgation of highly personal, private" forms of consumption (Graeber 2004: 74). Such an examination is revealing in the ways that a capitalism founded upon mass consumerism paradoxically depends upon encouraging certain forms of liberated decadent overconsumption. But to accomplish this may require the prohibition of unprofitable alternatives until they can be appropriated.

Progressive governmental sources and popular press cited the particular character of new immigrant enjoyment and desire as having its foundation in racial biology. This literature abounds with descriptions of how leisure time is spent by the foreign-born, often unproductively at the saloon or at festivities of a religious or celebratory nature. The many different religious holidays celebrated by different ethnic groups, which frequently interrupted work routines, are cited as barriers to productivity. Peter Roberts, a Progressive who wrote extensively on labor-capital relations in the anthracite region in the early twentieth century (1901, 1904, 1905), reveals his teetotaler agenda when he admonishes:

> The Sclav in his fatherland is predisposed to intemperance, but on new soil, where wages are high and social restraints few, where the forward meet and the baser passions are let loose . . . the Sclav's love for intoxication passes all bounds and the ruin wrought among them is great. (1905: 220)

In this passage Roberts suggested that the combination of relative wealth and freedom encountered by Eastern Europeans in America was squandered by their basal weakness.

Of Italians, an inclination towards frivolity and excitability was cited, as was their tendency toward criminality. Drawing upon the work of Italian criminal psychologists, American caricatures of Italian racial character distinguished between regional origins, describing the northern Italians as "cool, deliberate, patient, practical," and southerners as "excitable, impulsive, impracticable; as an individualist having little adaptability to highly organized society" (Dillingham Commission 1911c: 83). Early newspaper references to Italians in Hazleton lightheartedly played upon Italian caricatures, providing moralizing lessons on how to interpret the racial behavior of new residents. In a section of the paper reserved for urban news snippets smacking of boosterism, the belief in the coarse nature of Italian amusement was frequently caricatured, as, for example, in the *Hazleton Daily Sentinel* (July 31, 1879): "An Italian organ grinder ground out music along the streets last evening and drew together the usual crowd of gamins." Another short item reinforced racial hierarchy in the region, suggesting that, while "'the Germans turn naturally to music' . . . it runs in our minds that the Italians most frequently 'turn' to music" (*Hazleton Daily Sentinel*, August 23, 1879). These early characterizations persisted even when the Italian presence in the town amounted to more than a peripheral dimension of the local economy. During the 1900 strike in Hazleton, Italians workers were criticized for belittling the situation with their "jollification":

> The idle men residing in "Little Italy" look upon the strike as a sort of jollification, and yesterday there was an unusual amount of merriment. In many houses there were liquid refreshments and a good time was had by all. (*Hazleton Daily Sentinel*, September 22, 1900)

From these didactic images presented in newspapers, Hazletonians formed racialized interpretations of new immigrant lifestyles. In turn, the materiality differentiating citizens and foreigners manifested in landscape, material culture, and everyday social encounter was reinforced. Consumption was one such realm.

Consumption Habits of New Immigrants

New immigrant consumption practices had different implications or consequences for different parties. So long as employees continued to shop at the company store for certain necessities and work supplies, it was not of immediate or explicit concern for coal operators. If anything, the degree to which employees practiced thrift and self-reliance corresponded with the extent to which they would tolerate the meager wages they were offered. For Progressive reformers, whose concern was with the wage competition or assimilation of new immigrants, their consumption practices implied habits at odds with the economic and social stability of the dawning age of industrial democracy.

Economist and labor historian Frank Julian Warne, author of *The Slav Invasion and the Mine Workers* (1904a), considered consumption a labor issue by the degree to which Slavic thrift granted an unfair competitive advantage to wage earners capable of surviving on less than nativized workers. Like other early twentieth-century politicians, scholars, and social workers, these observers pinned the enfranchisement of America's new immigrants from Eastern and Southern Europe to their adoption of mass consumer habits (Cohen 1986). Importantly, these observers were not apprehensive about the satisfaction of immigrants' creaturely needs. Rather, their concern was with the kind of surplus desire that is patently unappeasable, what one observer of the Anthracite Coal Region (Roberts 1905: 220) defined as the "margin between [the] real wants and felt wants." In effect, a disregard for the normative desires of naturalized citizens amounted to a kind of threat.

He described an eviction scene of a Scottish family from a company town in the vicinity of Hazleton:

> All their effects on this eviction day were piled along the highway, a bureau, "straw ticks," a stove, several chairs, a rag carpet or two, with here and there a lithograph scattered incongruously among boxes, kitchen utensils, and the separated parts of beds. These and other belongings . . . of such quantity and cumbersomeness as to make necessary the employment of a wagon with horse and driver to remove them. (Warne 1904b: 224)

To this he compares the subsequent replacement in the house by a Slavic family who carried their meager belongings in a manner that suggested to Warne a "beast-of-burden adaptability to the most exacting physical labor" (1904b: 224). He describes their belongings:

> The cooking utensils of the newcomers were of the barest in quantity and quality. They had neither chairs nor bureaus. Their meager supply of clothing was all but limited to the garments they wore. "Straw ticks" and beds were conspicuous by their absence, the new occupants being content with rolling themselves in blankets and sleeping upon the uncarpeted floor.

Roberts, whose concern was of a social or moral character, spoke optimistically of the inevitable adoption by young Sclavs of American fashion, citing this as far preferable an expenditure to that of the saloon. Of the older generation's thrift, he suggested that "the pressure of increased wants must be brought to bear." He continued:

> The margin between their real wants and felt wants in clothing, rent and household furniture is very slight, and not until they are educated to place more value on an overcoat than "bock," on comfortable homes than beer dens, on good food than poor whiskey, will there be hope of their social advancement. (1904b: 220)

Even for observers in the anthracite region where the baseline of relative poverty remained prohibitively low, consumption practices held symbolic significance beyond providing a mere index of financial circumstance. Consumption practices indicated otherness, and their guided reform was therefore a solution to the problems arising from the tense interactions between native populations and foreign bodies.

Several key concerns echoed throughout the country by administrators, capitalists, progressive reformers, newspaper commentators, and social scientists wherever the new regimes of industrial capitalism attracted large amounts of immigrant labor. The challenge of American pluralism, the inequities that led to or inhibited interethnic or class struggle, and the dangers, moral and physical, presented to the native population by the presence of so many unassimilated foreigners taxed the imaginations of many at the time. Following the First World War, a kind of consensus developed among capitalists, advertisers, propagandists, and government administrators on a solution to these issues in the form of mass consumerism informed by beliefs in the pliable behavior of mass bodies.

THE ARCHAEOLOGY OF CONSUMERISM IN LATTIMER, PENNSYLVANIA

By the second quarter of the twentieth century, the materiality of the shanty enclave in Lattimer No. 2 began a rapid transformation. Slowly, it came to resemble a typical American suburb in architecture and spatial distribution. Those characteristics that once differentiated it from the other company homes were downplayed or erased, as were its origins as an ephemeral settlement of the marginalized. Simultaneously, consumer goods and other items used by its occupants began to reflect the characteristics of the mass consumer democracy being instituted throughout the country.

Beginning in the 1940s, the Lattimer Coal Company divested itself of ownership of the residential areas of Lattimer Nos. 1 and 2. Occupants, who previously paid ground rent while owning the structures erected on each plot, were given the option to move their homes or purchase the grounds. In about 1959 the house on Church Street site in Pardeesville was transferred to a new owner and at this time the privy in the backyard was filled by the contents of the house in a single clearance event, sealed with rubble, and abandoned. Artifacts were, for the most part, deposited in this single depositional event, divided into five strata of coal ash, rubble, and domestic and personal items, and capped with a layer of yard soil.

During the 2013 excavations, the privy was excavated and the contents catalogued. The assemblage included items such as machine-made, food-related bottles and decorated and undecorated tablewares.

It also included items reflecting new trends in consumerism such as personal objects made of plastic, for example, pens, jewelry, electrical appliances, and other accoutrements. Bottles and other containers were associated with commercial cosmetics and over-the-counter pharmaceuticals. Appliance dials and radio vacuum tubes reflected the residents' engagement with new forms of electronic media. In brief, the assemblage reflected the privy occupants' entanglement with the enticements and excesses of mass production and consumption and the enunciation of subjectivities inextricably networked to other broad-scale transformations in labor and manufacturing processes, infrastructure, commodity exchange networks, and the new discursive patterns of emerging electronic media.

To illuminate the many historical threads of this assemblage, we can use a theoretical framework developed from Marx's reflections on machines by the autonomist school of Italian neo-Marxism, what Felix Guattari and Gilles Deleuze describe as a "machinic system" or "machinic enslavement" (Deleuze and Guattari 2009; Guattari 2009; Lazzarato 2014; Raunig 2016). The development of *machinic mass consumerism* can be historicized only through a theoretical framework, which reveals the multitudinous nature of its material connections to infrastructures, subjectivation, the developments of science and technology, media, soft forms of coercion, and alienation.

MACHINIC FORMS OF PRODUCTION AND CONSUMPTION

Consumer behavior, interpreted through discourse-centered or semiotic models adapted from poststructuralism, is often considered to be intelligible as an enunciation of shared symbolic, ideological, and identifying systems. Such accounts focus on the practices of consumer behavior or household economy in isolation from the manner in which structures of production are proliferated in commodity exchange.[1] Machinic theory, however, suggests that consumerism cannot be understood simply as a matter of knowledge, consciousness, and discourse, but rather is composed of a heterogeneous assemblage of unconscious and material processes, ecologies, and infrastructures.

Machinic theory is derived from Marx's musings on the role of technical machines in the extraction of surplus value from producers in industrial production processes. Beyond merely the replacement of a worker with a machine, Marx recognized that the *machinic* entails the articulations of many things:

> an automatic system of machinery . . . set in motion by an automaton, a moving power that moves itself; this automaton consisting of numerous mechanical and intellectual organs, so that the workers themselves are cast merely as its conscious linkages. (Marx 1972: 132)

From Marx's musings on machines, scholars of the Italian Autonomist or post-Operaist school and others developed the concept of machinic systems as means to understand broader capitalist systems of profit capture (Deleuze and Guattari 2009; Guattari 2009, 2011; Lazzarato 2014; Raunig 2010, 2016). Recognizing that the factory is only one setting for the actualization of the systematizing and articulating capacities of capitalism, they pursue these processes as analogical ways of understanding broader heterogeneous systems of commodity production, labor exploitation, and subjectivation. As such, technical machines are only one subset of a system articulating social, environmental, and biological components. Recognizing the nodes and linkages within a machinic network requires that we transcend traditional dualisms between micro and macro processes, histories and presents, knowledge and materials, humans and machines, and material and ideological processes.

An important intervention enabled by the recognition of the machine-like articulations of this heterogeneous network is that the separation of production, labor, and consumption in the commodity cycle of capitalism is rendered as merely an arbitrary, ideological, or rhetorical division (Roller 2015; Wurst and McGuire 1999, Wurst 2015). But even this intervention does not go far enough. What must be understood is that in between these two nodes—which happen only to be the ones most conspicuously populated by everyday human activities and amplified and annotated by a superabundance of discursive or semiotic content—are other equally emphatic connective segments. These include passive material elements of the capitalist ecology such as advanced mass transit, the distribution networks of chain stores, the proliferation of advertising, and developments in materials science. Each of these developments, which are both technical and social, serve to reinforce the stability and effectiveness of a totalized economic system based upon steadily accelerating processes of production and consumption.

The goal of machinic theory is not simply to produce a rich descriptive analysis of the heterogeneous elements of a network. Rather, Marx's humanistic analysis of machines was founded on a mandate to discern the sources of alienation and violence to people that materialize in their deployment (Sayers 2014, 2003). In this way, the key commodity conceived, developed, and manufactured by capitalism can be understood as consisting of the subjectivity of humans developed in and among systems of machinic enslavement (Lazzarato 2014; Matthews 2010).

It is in the Interwar Period or the Machine Age that a significant rupture in the very materiality of everyday life occurs from what came before. This change is noticeable in the archaeological record as the uncanny reflection of our present material world, the moment when artifacts recovered appear more like those of the present than of any moment in the distant past. In this period, the stage is set for new regimes of political and material economy: exponentially productive machine-based industrialization; sociologically informed workplace control; politically integrated labor organization; Fordist and then post-Fordist consumer-producer economies; logistical, infrastructural, and technological innovation; and a seemingly infinite capacity to develop new consumer markets through a variety of means.

MACHINIC CONSUMERISM AND THE INTERWAR PERIOD: THE DUAL CRISIS OF POLITICS AND ECONOMY, SOCIAL INSTABILITY, AND SURPLUS PRODUCTION

Machine Age mass production represents the reversal of the central principle of previous economic models. Whereas previous models depended upon the selling of "a few things to the richest people," it now offered to sell "great quantities of things to great masses of people" (Filene 1930: 8). The architects of this change, including businessmen, designers, advertisers, producers, and others, encouraged the production of commodities attuned to the particular assumptions they held about the nature of their market and the infrastructure that produced and distributed it.

The development of mass consumerist practices in the early twentieth century was not solely the product of capitalist impulse. Rather, it served as a diabolically incisive solution to two major crises confronting the nation's elites at the turn of the twentieth century. By the beginning of the twentieth century, industry had expanded production far beyond the absorbent capacities of its market through the adoption of mechanization, advancements in science and technology, and rationalized administrative practices such as scientific labor management. Additionally, decades of mass migration entering the country throughout the late nineteenth century bolstered the ranks of the low-wage, unskilled labor force required by these new production regimes. The glut of commodities produced by these material factors required some form of market expansion, either as an extension into new territories (such as colonization can produce) or a transformation (or "deepening") of current markets (Paynter 1988: 418). The introduction of Fordism in 1910, comprising an expansion and stabilization of workers' income and leisure time, pursued the institution of the latter effort. However, this quantitative increase in wages and leisure time was insufficient to bring about a needed qualitative change in the nature of consumer desire (Ewen 1976: 24). The government and business community recognized that the problem of underconsumption was fundamentally a social one.

The other crisis was political. Around the turn of the century, middle-class observers began to grow fearful of the combined threat of labor unrest, foreign immigration, and mounting inequality as a force that would destabilize the precarious social order. For some observers, the greatest fear came from the possibility that radical labor might organize the unskilled immigrant workers filling industrial roles throughout the country.

Efforts to confront these crises found both rationale and method from unlikely intellectual sources. Ideas about the psychology of masses, crowds, and publics adapted from sociology and psychoanalysis furnished new ways to conceive of the relationships between individuals, objects, desires, and social bodies. Starting in the mid- to late nineteenth century, European and American scholars of sociology, philosophy, and psychology developed theoretical conceptions about the behavior of social bodies. In the process they captured the attention of middle-class anxieties about social instability and the incendiary capacity of crowds and other social aggregates.

Throughout the turbulent period of class struggle and political unrest leading up to the *fin de siècle*, European sociologists such as Gustave Le Bon and Gabriel Tarde developed influential theories about crowd dynamics. Le Bon conceived of the crowd as forming into a single organismic social body governed by unconscious rules, motivations, and actions at the scale of both individual and aggregate. At once inherently empty and suggestible, he believed the crowd developed a volatile collective unconsciousness with its own properties, absorbing ideas by a process without reason. "Tolerat[ing] neither discussion nor contradiction," le Bon contended, ". . . . the suggestions brought to bear on them invade the entire field of their understanding and tend at once to transform themselves into acts" (Le Bon 1895: 43).

These ideas would be readily applied by the propaganda campaign instituted with the beginning of American involvement in the First World War. Government and industry moved quickly to neutralize threats to American industry and the war effort, focusing special attention on the support of workers in steel, textiles, mining, and other vital industries. On April 13, 1917, President Wilson created the Committee for Public Information (CPI) to develop public support for the war using the techniques of propaganda. He merged the findings of mass psychology with the methods of Progressive journalism, public relations, and consumer advertising—a potent mix. These ideas were applied through every communicative form available at the time. Recognizing the need to reach a direct rapport with individuals from all walks of life, the committee endeavored to "bypass established social structures" to get directly at crowds (Leach 1986: 103–104). Propaganda efforts entered movie houses, schools, and every article of popular and industrial literature. Great effort was made to penetrate isolated immigrant industrial communities such as Lattimer, which had the power and motivation to resist support for the war effort (Dubofsky 1996: 133–146).

Following the war, the work of the propaganda campaign was revisited, refined, and institutionalized by industries and intellectuals as an unexpectedly successful model of scientific and theoretical approaches to managing the public (Bernays 1928: 27; Ewen 1996: 146–173; Leach 1992). Edward Bernays, a veteran of the CPI, nephew of Sigmund

Freud, and founder of the school of public relations, writes (1928: 27–28):

> It was. . . . the astounding success of propaganda during the war that opened the eyes of the intelligent few in all departments of life to the possibilities of regimenting the public mind. . . . The manipulators of patriotic opinion made use of the mental clichés and emotional habits of the public to produce mass reactions against the alleged atrocities, the terror and the tyranny of the enemy. It was only natural, after the war ended, that intelligent persons should ask themselves whether it was not possible to apply a similar technique to the problems of peace.

These ideas were disseminated to universities, company offices, and government agencies through manifestos, textbooks, and instructional manuals. The terms *public* and *masses* are used interchangeably in these texts, defining corresponding properties of the abstracted interconnectedness and pliability inferred from the "dispersed and passive crowd of uprooted individuals" making up the American population (Leach 1986: 100). Read against the grain, they revealed striking assumptions about the nature and relationship of individuals and government, society and economy. Three ideas unite these texts:

- The public is irrational, self-interested, and vulnerable to fragmentation and disorder without the guidance of informed elites (Bernays 1928; Ewen 1976; Curtis 2002).
- An American society, fragmented by cleavages of class, race, and political persuasion, can be united by a shared participation in consumption and/or a regulated engagement with an "engineered" consensual public opinion (Bernays 1923, 1928; Ewen 1976: 188–189; Hickel and Khan 2012).
- To keep up with surplus production, the American economy needs to create consumer desire where there is none and psychological techniques can be applied to successfully accomplish this (Ewen 1988; Filene 1930, 1934; Sheldon and Arens 1932).

The first two decades of the twentieth century saw a flurry of federal interventions aimed at stabilizing American markets, neutralizing dissent, and engineering social demographics. Coercive measures to suppress dissent appeared in the form of legal frameworks such as the Espionage and Sedition Acts of 1917 and 1918 (Ewen 1996: 119–121). Immigration reform laws included the Anarchist Exclusion Act of 1901 and the passing of the Immigration Act of 1924 following the analysis of the Dillingham Commission (Ngai 1999, 2004a, 2004b; Zeidel 2004). Of the traditional methods of controlling populations through physical coercion, however, Bernays felt his alternative methods were more effective. Quoting German biographer Emil Ludwig, he cites the "impotence of force to organize anything" (Bernays 1928: 18). Rather, through knowledge of predictable mechanistic responses, individuals can be manipulated to "create public acceptance for a particular idea or commodity" (Bernays 1928: 18).

THE ARCHAEOLOGY OF MACHINE-AGE POLITICAL ECONOMY: THE IMPRESSION OF MACHINES

The mid-twentieth century artifact assemblage recovered from privy and yard soils at Church Street in Lattimer No. 2 is an expression of individual and household consumer choices. However, when seen within the *longue durée*, these choices also reflect the structures of historical contingency: an engineered trajectory structuring household practice to be in alignment with the goals of Fordist and post-Fordist finance and governance. The adoption of machinic consumerism can be identified through at least seven characteristics identified in the assemblage recovered from the Simone Lot on Church Street in Pardeesville, Pennsylvania. These aspects, discussed in detail below, include: (1) Mechanization and Mass Production, (2) Logistics and Infrastructure, (3) Chain Stores and Mass Distribution, (4) Advertisement, (5) Product Design and the Fashion Cycle, (6) Obsoletism and Waste, and (7) Media.

Mechanization and Mass Production

When we encounter labor struggle throughout history, it attains to a monadic quality. Although it often reads as a collection of particularist struggles—by shop floor, by industry, by region, by interest group, by union—there is cohesiveness across each narrative. As Walter Benjamin might say, each struggle is a crystallization of the broader class struggle itself. In the consumption practices and the very materiality of commodities from the 1930s and 1940s excavated from the shanties of ethnic laborers and their families, we recognize the interpenetration of multiple sites of struggle within a broader system of capitalism in constant transformation.

In "Steps to an Archaeology of Capitalism" (1988), Robert Paynter suggests that the morphology of items such as glassware and ceramics, used by archaeologists as indicators of diagnostic time or socioeconomic status, is rarely recognized as reflecting the introduction of social ideologies that defined the roles and positions of labor and class in society. Mechanization implied not only the introduction of machines into the processes of assembly and extraction, but a reorganizing, disciplining, and de-skilling of human labor. Mechanized processes reorder the relationship of laborers to products by extracting the accumulated skill and knowledge of workers and endowing machines with these attributes. In the process, skilled craft is rendered obsolete. Labor takes on the character of the tending of machines, circumscribed to a rationalized fragment along a *chaîne opératoire*, or a machinic network of assembly line production. In such a context, enduring the slow-motion shock of repetition, monotony, and time discipline took on the character of the workers' emotional and physical exchange for wages (Buck-Morss 1992; Ewen 1976: 27). Paynter asserts that the introduction of machining processes into bottle making is connected to efforts by management to replace or otherwise disempower labor resistance by replacing skilled workers with the fixed capital of machines (1988: 420–421). Commodity fetishism, in turn, obscured these efforts, separating production from consumption, while drawing "new relations between goods and people" (Trachtenberg 1982: 133).

Starting in this period, the majority of the objects in the assemblage shows indications of mechanized production. Glass bottles exhibit characteristics of semiautomatic processes such as Owens suction scars (SHA 2014). They also include many constructed by fully automatic production from the 1940s. These bottles include base knurling and stippling that not only provided traction for conveyor belt production, but also served to conceal the signs of production on the finished product (SHA 2014). In short, the aesthetics of machine-made bottles served to alienate consumers from producers. Ultimately, systems of mass production channel workers not only to sell their labor, but to hire the labor of remote others at the going rates to produce the everyday items they consume. In a sense, this system makes all laborers, coal miners, glassmakers, textile workers, and so on, complicit with the reigning labor regime. Within the broader scales of political economy described above, this had implications for defining their own class positions within mass society.

Logistics and Infrastructure: Permissive Technologies and the Industrial Landscape

Capitalism's capacity to extract profit from commodity production demands the persistent annihilation of time through increasing control over space (Harvey 1989). As the velocity of capital flow around the commodity cycle increases, rates of profit in turn may increase. In the present age, the mastery of space and time has been developed into a science known as *logistics*, adapted out of wartime strategies (Angelil and Siress 2011; Lyster 2012). Today, transnational corporations such as Amazon, Wal-Mart, and FedEx have mastered time and space as their central product, producing virtually nothing but the capacity to efficiently concentrate, transport, and redirect commodities across global markets. This economic sector depends upon a particular materiality assembled out of transportation networks, infrastructure, and communication technologies. It represents a machinization of the exchange and distribution networks in correspondence with alterations made of the factory floor beginning nearly a century ago. In essence, as machines increased the velocity of factory production, the technological and social network of distribution, exchange, and consumption expanded to meet these new velocities and spatialities. These efforts entailed a wholesale reordering of the landscape according to the needs and capabilities of capital and the state.

Intensified development of this landscape began in the interwar period as technologies and infrastructure developed to exponentially meet the productive capabilities of machine-age production, coupled with monetary policies conducive to unbridled growth. By the end of the Second World War, with state coffers flush from postwar economy, the federal government began to invest in this development, funding the construction of highways, rail lines, airports, landscapes, and utilities (Harvey 1989: 133–135).

Described as "permissive technologies" by Bluestone and Harrison (1982: 115), infrastructures that permit the radical mobility of capital, technology, information, and commodities meant companies were no longer bound to industrial centers. For laborers and producers, this meant an added precarity to regional economies as companies could move entire industrial operations at will, loading machines onto railcars or cargo jets. First in the 1930s, and again in the 1960s and 1970s, the practice of exporting the sites of production, described in 1930s parlance as "runaway shops," was an often-used tactic to abandon labor markets whose collectivization hampered profit margins (Bluestone and Harrison 1982; Cowie 1999).

Symptoms of permissive technologies that were manifested in logistics and infrastructure can be seen in the archaeological record if an assemblage is observed in the spatial context of both a regional scale and the national and global. Although the anthracite coal industry dominated the economy of the Hazleton region, jobs in the manufacturing of textiles, clothing, shoes, food products, buttons, and bottling provided revenue for the cash-strapped families of miners. Employing surplus labor in the region including men, women, and children, this light manufacturing arguably provided a relatively unalienated spatial relationship to available consumer goods. While forms of factory discipline undoubtedly presided across these local industries, consumer goods purchased, used, and shared might have been associated with the labor of extended families and came bearing the manufacturing location of familiar places. Hazleton alone contained at least eight breweries during the last quarter of the nineteenth century. Oral histories conducted with residents from Lattimer and elsewhere document memories of lining up at the Hazleton Soda Factory to collect free "seconds." Interviews also document memories of uncles and cousins employed in these industries.

Maker's marks and other manufacturing inscriptions on consumer goods, including beer, soda, and milk bottles as well as rubber buttons from the first half of the twentieth century, indicate a strong representation of local industries in certain sectors of the consumer market. By the beginning of the second half of the twentieth century, however, many manufacturing industries had closed or relocated to centralized manufacturing centers. These included all local breweries and bottlers. Locational information coded on machine-made bottles document that these and other industries concentrated elsewhere in the country, facilitated by an infrastructure of mass transport and communication.

With their labor, the producers of coal yielded the "mucky material bottom" underwriting the energy, light, and movement of modernization and mechanization (Arnold 2014: 266–267). But this mobility enabled the dispersal of forms of alienation far greater than what occurred on factory floors and company towns across the country. It mystifies the origins of commodities by concealing the everyday realities of their production, including the living and working conditions of their producers, from all consumers.

Chain Stores and Mass Distribution

In the *Arcades Project,* Benjamin astutely observed that, "at the moment when the production process closes itself off to people, the stock in trade becomes accessible to them—in the form of the department store" (1999: 367). The centralization of administration and production accommodated by the extension of permissive technologies also required an in-kind alteration of the sites of exchange. As documented by Mullins (1999) and Cohen (1990), beginning in the 1930s chain stores spread through urban neighborhoods, challenging regional dynamics of commerce operating on the scale of the corner store. Mullins and Cohen argue that ethnic and racial politics were often played out in different ways at the local level, before and after these changes. An example comes out of the case of African Americans in Annapolis in mid-century who negotiated racism by preferencing shopping environments and products associated with brand-name goods. Mullins suggests that identifying the multivalence of specific consumer groups' tactical negotiation of the coded languages of racialized or politicized mass consumerism entails "situational negotiation, subtle tactics, persistent aspiration and constant circuitousness," which can be difficult to observe in the archaeological record (Mullins 1999: 173, 187–189).

The material signature I am identifying here is not one that precludes or negates the empirical validity of these practices, but that seeks to contextualize these practices into the broader network of givens, coerced and influenced desires, urgencies, dominations, challenges, and normative persuasions. Cohen suggests that by the end of the 1920s, mass-produced brand-name goods sold in chain stores made up half of the goods consumed in major cities across the country, even in the ethnic neighborhoods of new immigrants for whom neighborhood commerce played a significant role in maintaining social solidarity (1990: 107). For scholarship such as Cohen's, rooted in the class and ethnic identitarian analytical scales of the new social history, the rise of Golden Age mass consumerism marks the end of an era for particularistic studies of this kind. Rather, the focus of consumerism scholarship from this historical moment forward demands an approach both to global or national networks and to the mass social production of desire.

Advertisement: Consumption as Social Production

Raymond Williams proposes that the forms of advertisement developed in the interwar period are "a magic system . . . a highly organized and professional system of magical inducements and satisfactions . . . strangely coexistent with a highly developed scientific technology" (1993: 422). In the context of mass consumerism, advertising is, first and foremost, a form of social production (Ewen 1976: 23–41), which depends upon the decoupling or obscuring of the relationship between commodities and the specificities of the productive process. As manufacturers recognized that the social production of new markets was essential to corporate survival in the age of mass production, knowledge of customer attitudes and techniques for their manipulation became equivalent in value to technological innovations or scientific applications (Ewen 1976: 26; Trachtenberg 1982: 135–137). The growth of professionals devoted to these services resulted in extra costs assumed as part of the production process. Advertising joined other aspects of fixed capital along with labor and things like the maintenance and operation of machinery.

Whereas in previous centuries, advertising more or less applied the powers of persuasion directly to the prospective customers, by the interwar period it appealed indirectly to "basic personal relationships and anxieties" (Williams 1993: 418). Edward Bernays, in his book *Propaganda* (1928: 52–53), is explicit in his application of the lessons of psychoanalysis to a sophisticated advertising methodology that owes as much to Freud as to Marx:

> It is chiefly the psychologists of the school of Freud who have pointed out that many of man's thoughts and actions are compensatory substitutes for desires which he has been obliged to suppress. A thing may be desired not for its intrinsic worth or usefulness, but because he has unconsciously come to see in it a symbol of something else, the desire for which he is ashamed to admit to himself. . . . This general principle, that men are very largely actuated by motives which they conceal from themselves, is as true of mass as of individual psychology. . . . Human desires are the steam which makes the social machine work. Only by understanding them can the propagandist control that vast, loose-jointed mechanism which is modern society.

The presence of many name-brand items related to grooming and health, the fastest-growing sector of mass consumer industry, suggests the influence of advertisement in the consumer habits of the occupants of Pardeesville by the interwar period (Ewen 1976; Williams 1993). These items include vitamin and pill bottles, anesthetic cream, toothpaste tubes, toiletries, makeup, and other accoutrements (Figure 5.2).

Stuart Ewen (1976) argues that interwar consumerism cultivated youth as a cultural symbol of renewal, beauty, and progress. This ideology served a double role as social production. First, it served to establish "the period of childhood and adolescence into a period of consumption" as a new market for products (1976: 139). Concealed in this ideological construct, however, is a strategic intervention in the nature of authority and tradition as they are connected to productive labor. Mechanizing labor processes replaced demand for developed knowledge and skills held by elder craftsman with the unskilled tending of machines. For these processes, strength, endurance, and malleability were prized in the workplace. Product marketing celebrated youth, or the product-aided return to a youthful condition of vitality, energy, and beauty. Cosmetics, creams, and soaps promised this in advertisements featuring youth while targeting the insecurities of all ages. The same can be said for the great variety of mass-produced pain-numbing medicines sold at the time that functioned as anesthesia to maintain the productivity of laborers despite the inevitable pain accompanying repetitive motion (Buck-Morss 1992).

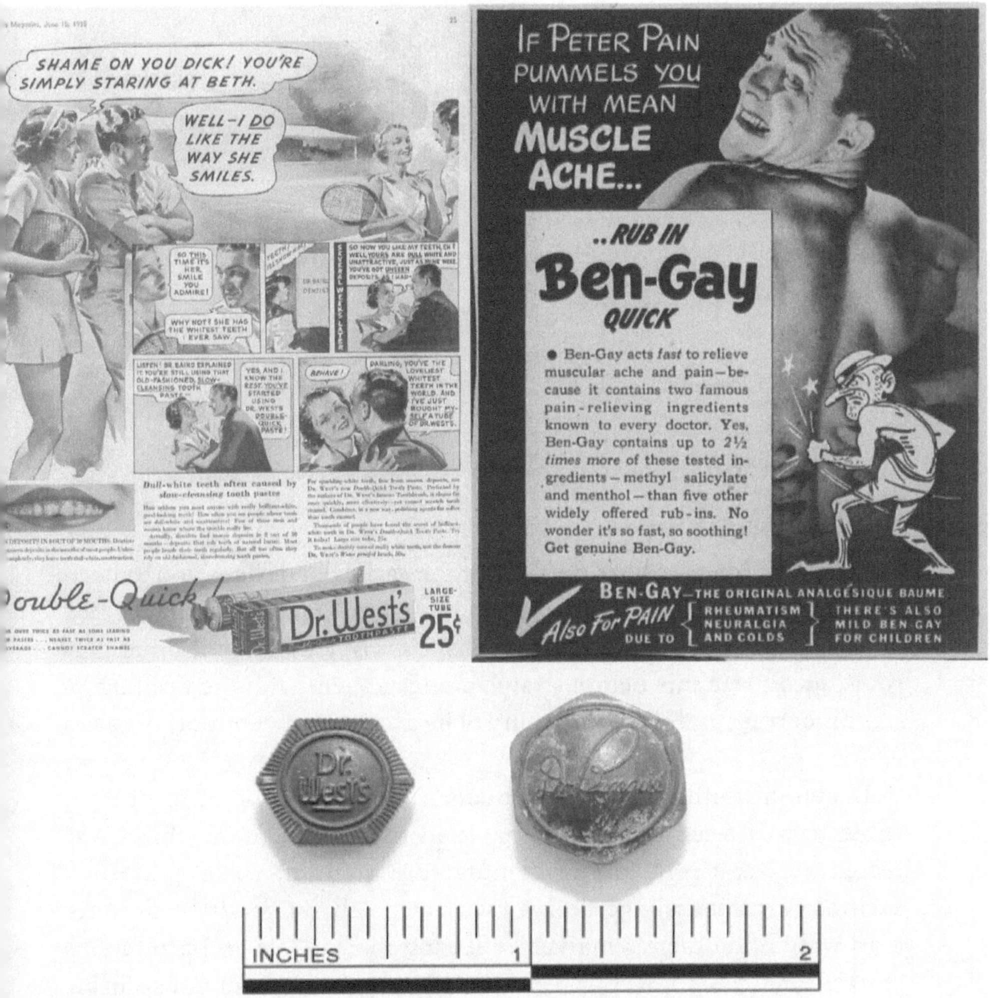

Figure 5.2. Dr. West's Toothpaste cap and ad; Ben-Gay analgesic balm tube cap and ad from 1931 (images from Duke University Digital Archives).

Product Design and the Fashion Cycle

The plasticity of artificial materials opened up entirely new dimensions of aesthetic desire. Roland Barthes (1988: 92) said of plastic: "It is the very idea of its infinite transformation; it is ubiquity made visible. Plastic remains impregnated throughout with this wonder: it is less a thing than the trace of a movement." The Pardeesville assemblage

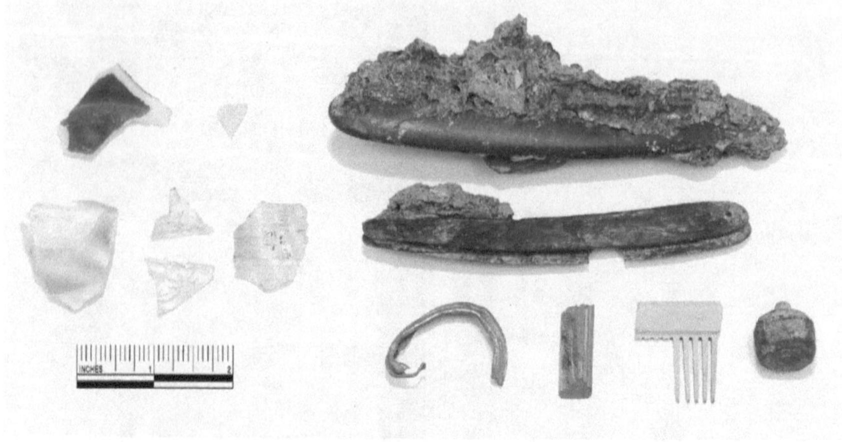

Figure 5.3. Artifacts representing new decorative technologies (photo by author).

contains not only artifacts notable for the variety of designs and materials but also detached ornamentation of chrome and plastic. Recovered objects such as appliance knobs, pens, combs, jewelry, and other personal accoutrements demonstrate an engagement with the fashions of late modernity and the temporality of its accelerated commodity cycles (Figure 5.3).

Beginning with industrial production, the separation of form from substance as a manipulable variable in the design of an object was recognized as a principle of manufacturing (Ewen 1988: 33). However, early industrial-age design focused mostly on goods secured for a growing middle-class market, reflecting their values and aspirations (Ewen 1988: 32–40; Wall 1994). Recognizing the necessity of securing mass consumer patronage for mass-produced goods, manufacturers increasingly focused on product design as the key to stimulating mass consumption.

Psychoanalytic insights contributed to the development of wish-fulfilling aesthetics as enticement for customers to purchase machine-age products (Ewen 1988). Roy Sheldon and Egmont Aren's 1932 design manual, *Consumer Engineering,* instructed generations of designers in "the exploitation of the 'sublimated sense' in the field of product design" (1932, quoted in Ewen 1988: 49–50). They argue that haptic connections to material objects often serve as the deciding factor in decisions to consume. Recognizing its effect over quality and craftsmanship, style emerged as a way to heighten the fetishistic qualities of cheaply produced goods. In the process, they further obfuscated their physical relationship to production.

New technologies and scientific knowledge provided freedom for manufacturers to manipulate surfaces, materials, and forms in ways that confounded the senses but still referenced aesthetic history. Pressed forms and iridized, baked-on colors on glass vessels could be produced cheaply, mixing modern surfaces with historical forms. Technological processes and materials that imitated natural or handcrafted materials and forms including patent leather, poured concrete, chromed metal, wood-grained Bakelite, plastic costume jewelry, electroplated silver and gold, and stamped, decaled, and pressed forms all challenged a class-based hierarchy of aesthetics and materials with a democracy of consumer forms (Barthes 1988; Mrozowski 2006: 135; Schwartz 1996: 192–196). By the late 1930s, plastics had yielded such a diversity of qualities that few functional forms escaped its embodiment.

Obsoletism and Waste

The desires implicit behind accelerated consumption demanded a corollary, new practice of waste and replacement that would ensure a constant turnover of demand. Aestheticizing the formal aspects of utilitarian objects opened them up to increased capitalist control over the temporality of the fashion cycle (Ewen 1988: 233–258). Designers and advertisers, applying psychological techniques, focused their attention not on successive improvements but rather on the engineering of persistent dissatisfaction. Earnest Elmo Calkins (1930, quoted in Ewen 1988: 243) described the goals of obsoletism as an effort

> to make the customer discontented with his old type of fountain pen, kitchen utensil, bathroom or motor car, because it is old fashioned, out of date. . . . We no longer wait for things to wear out. We displace them with others that are not more effective but more attractive.

Product designers strategically eliminated anachronistic styles to pursue new “forward-looking” aesthetics, combining new forms with modern materials such as the great variety of plastics developed in the 1930s. Consolidated control of advertisement, product design, and the sensorium of chain retail spaces gave manufacturers further control over the temporality of consumption and waste on a level never before encountered.

Diagnostic materials from between about 1920 and 1940 make up the majority of the total identifiable assemblage at Lattimer. This assemblage is striking both in the quantity and diversity of materials and forms. Disposable glass containers, appliance fragments, toys, and personal items such as pens, perfume bottles, combs, light bulbs, and glassware make up much of this bulk. From earlier eras, item types are limited to tobacco pipes, ceramic fragments, nails, and container glass. Deposition, disposal practices, and general poverty account for some of this incongruity, but the sheer quantity and diversity of the post-1920 assemblage can also be understood in the context of the condition of materials: discarded complete, *en masse*, or partly consumed. These attributes point toward emergent discard practices connected to new relationships between mass-produced objects and consumers.

Media as Social Production

Radio vacuum tubes, fragments of 78 rpm records, and appliance dials recovered from the assemblage at Lattimer demonstrate that even occupants of shanty settlements were deeply involved in new mass media formats emergent in this period. The proliferation of radio sets serves as a double form of consumption. First, through the acquisition of the physical components, but more importantly, for the present discussion, radio programming served as a vector for the entanglement of consumers in an ideological world of collective desire, fear, and imagination. Moreover, sociologists recognized that the collective energies unleashed by electronic media bypassed the dangers of crowd dynamics touted by early mass psychologists. By early in the twentieth century, media operated in capitalist society in ways that Marx could never have predicted without the imagination of science fiction.

Gabriel Tarde, a close associate of Le Bon, offered a solution to the problem of the fragmenting effects of collective emotion and instability inherent to crowd dynamics. Tarde elucidated means by which elites might take advantage of the suggestibility of groups, rechristening this new pliable mass aggregate "the public" (Tarde 1903, 1969). Observing that as lines of communication and transportation spread by newspaper, the telegraph, railroad, and steamship, the public increasingly drew opinions "no longer grounded in the immediacy of their lives" but from the abstract realm of a marketplace remote and yet essentially "coextensive with society itself" (Ewen 1996: 68; Tarde 1969).

Ideas from Le Bon, Tarde, and others were reinterpreted through the lens of the particular crisis entailed by early–twentieth-century American pluralism, modernity, and progress. Social psychologist Edward Ross took up this line of thought, asserting that "presence is not essential to mass suggestion. Mental touch is no longer bound up with physical proximity. . . . with [the telegraph and newspaper] remote people are brought, as it were, into one another's presence" (Ross 1908: 63). Ross and others believed that the crowd, dependent upon face-to-face contact for the spread of ideas, would be superseded in the next century by a public united and stabilized through the "space-annihilating" apparatus of new media (Ewen 1996: 70–73; Ross 1908; Susman 1973; Williams 1993). In the hands of these thinkers, the crowd's lack of stability and capacity for volatility became advantageous characteristics for social control when combined with the burgeoning promise of media networking (Leach 1986: 102–103, 105).

Between 1920 and 1930, radio ownership and usage grew in popularity among hobbyists and regional institutions. Regional programming from ethnic, religious, and political institutions flourished, and as early as 1922, local news, entertainment, and advertising were broadcast in 3 million homes (Ewen 1996). Recognizing the potential power of this new medium to reach the masses instantaneously by a central nervous system, radio wave usage was brought under government regulation and then corporate control with the passage of the Radio Act of 1927 and the formation of the FCC (Federal Radio Commission) (Willey and Rice 1933b: 195–196). The 1927 Act, passed as an emergency measure to "bring order" to the "chaos" of crowded unregulated frequency usage, paved the way for the allocation of radio frequencies to corporate monopolies by the sovereign power of the state. Of the physical sets themselves, sales of radios amounted to a profit of $852 million by 1929 (Mowry and Brownell 1981: 1–4). Beginning in the late 1920s, industry came to recognize the potential for massive profits from the manufacture of radio sets, components, and programming. A massive consolidation of companies ensued, resulting in the standardization of products into interchangeable components such as vacuum tubes and capacitors (Harvard Business School 2014).

Following the Great Depression, President Hoover coordinated a massive study of the habits, customs, and trends of the population, with an "emphasis on elements of instability." Reporting on radio ownership collected in the 1930 census, the Report of the President's Research Committee on Social Trends (Willey and Rice 1933a: 214) proposed that with an average of 3.1 listeners per radio set, radio could reach a total of 37,442,869 listeners by 1933. In their study of radio usage, sociologists Willey and Rice warned:

> It is as agencies of control that the newspaper, the motion picture and the radio raise problems of social importance. The brief survey of their development in each instance shows increased utilization coupled with concentration of facilities. . . . Mass impression on so vast a scale has never before been possible. The individual . . . increasingly utilizes these media and they inevitably modify his attitudes and behavior. What these modifications are to be depends entirely upon those who control the agencies. Greater possibilities for social manipulation, for ends that are selfish or socially desirable, have never existed. (Willey and Rice 1933a: 215)

Approaching the issue systematically in 1935, social psychologists Cantrill and Allport mapped the depth and reach of radio messaging, contrasting its effects with the rousing face-to-face stimulation of crowd phenomenon. They suggested that radio had "a slightly dulling effect upon higher mental processes" but cheerily announced it also attained a "standardizing influence" that can serve to "counteract fragmentation" (1935: 13, 140). By the period just after the Second World War, critical theorists of the Frankfurt School targeted media and mass entertainment as major producers of a culture of consumption that neutralized political dissent (Benjamin 1970; Adorno and Horkheimer 1944).

As objects recovered from archaeological sites from the early part of the century, electronic media do not simply represent a terminal point for physical commodities. Rather, they can be historicized as tracing the beginnings of a communicative network facilitating the consumption, negotiation, or appropriation of ideologies, imagined identities, and shared experiences. From these objects we can write the beginnings of an archaeology of electronic social media articulated with that of mass consumerism and industrial democracy.

DISCUSSION

In 1929 businessman Edward Filene proposed that mass production "holds possibilities of accomplishing for mankind all of the good that theoretical reformers or irrational radicals hope to secure by revolutionary means" (Filene 1929: 9, quoted in Ewen 1976: 87). For Filene and others, this would be accomplished by ameliorating, concealing, or distracting citizens from the material conditions that precipitate class tension. The concept of machinic mass production as wielded by elites such as Filene entailed more than simply the introduction of new industrial processes to factory settings or expanded integration, reach, and effectiveness of capitalist endeavors. It also entailed the engineered development of a machinic network linking consumerism, production, exchange, distribution, and disposal to an assemblage of political, social, material, and financial structures.

Mark Leone and Paul Shackel (1987) observe that in the Chesapeake region during the eighteenth-century, mass-produced goods such as toothbrushes, clocks, and table settings flooded the market, disciplining consumers into the regimen of hierarchy and segmentation embedded in mercantile capitalism. Shackel concludes that the class segmentation arising in this era can be connected to the Industrial Revolution that followed to the degree to which these "new social behaviors reinforced the lower classes' ties to their inferior social position." These same workers, Shackel concludes, would "eventually bec[o]me the laborers of the new industrial era" (1993:30). The mass consumer revolution of the early- to-mid–twentieth century was similarly inextricable from economic ideologies of the moment. However, they conform to the particular temporality, spatiality, and materiality of twentieth-century industrial landscapes.

Machinic mass consumerism, founded upon the production of new forms of consumer desire, served not only to capture excess profit but also to assemble subjects into vertically integrated formats, both politically and economically. It can be argued that the development of machinic mass consumerism had two major implications for all subjects: alienation from the broader political-social context of production, and an intensification of the relationship between personal desires and commodities. Both of these forms of alienation entail forms of memory loss implied by the process of reification as elaborated upon by Lukács (Connerton 2009: 43; Lukács 1971). For an object or other entity to successfully take on the character of a commodity opened up to receive the values of consumer desire, the historical processes that structure its creation need to be forgotten so that it can accrue the new values of exchange and profit (Connerton 2009: 43; Sayers 2014: 36–42). Reification implies the process whereby labor and the products that create the material world of capitalism are alienated from producers and the fundamental subjective realities of their everyday life. Instead, these objects, social relationships, and the processes wherein work is conducted take on an objective social quality that is distanced from an individual's life (Lukács 1971: 83–149; Marx 1978; Sayers 2014: 27–45). Moreover, mass consumerism isolated individual choice from ethical, moral, or collectively negotiated principles. Using psychoanalytic techniques, individuals were encouraged to freely exercise their selfish drives, provided they remain within the confines of a stabilized social structure. Retrospective reckoning suggests that this was accomplished less by the satisfaction of actual needs but rather the expansion, channeling, and sublimation of consumer greed, selfishness, and other drives. This amounts to the equivalent of a reversal of the lyrics of a famous song by the Rolling Stones, namely, "You can always get what you *want*, but you can't always get what you *need*."

That mass consumer practices show up so abundantly in the waste of the most excluded of producers, the occupants of a shanty enclave in the coal region, suggests something of the ubiquity of its material signature. For producers such as the mine laborers at Pardeesville, this change entailed a new emphasis upon their values as consumers. This reflects the manner in which machinic subjectivities may often be at odds within themselves as a panoply of new roles articulate with different nodes of the network. A single individual may occupy the tension of being simultaneously a Fordist worker, a pliant machine tender, a union adherent, a brand loyalist, a thrifty entrepreneur, a decadent squanderer, a self-reliant debtor, a radio listener, and a citizen. Machinic theory brings insight to the fact that these contradictory elements may not be reconcilable at a discursive level, but rather exclusively at a material or infrastructural level. As asserted by Guattari, "The ground of enunciation is most often existential, not discursive" (Lazzarato 2014: 65).

6

Migration, Part 2

Emancipation and Catastrophe

Time accomplishes for the poor what money does for the rich.
Cesar Chavez

In the preface, I briefly introduced Bernardo Bertolucci's 1976 film *Novecento,* about political consciousness and peasant revolt in Northern Italy. Rather than providing a progressive, linear recounting of history, Bertolucci structures the film in a way that is reflective of revolutionary modes of heritage and memory making. It is an interesting model to contemplate how the revolutionary spirit of the Lattimer uprising reappears repeatedly in the region's memory. Bertolucci's explicitly political filmmaking depicts moments of political consciousness and emancipatory action in the twentieth century, not as linear processes, but within a temporality that is cyclical, gestative, and self-reflexive. He depicts the revolutionary moment twice, first at the outset of the film as a flashback, and then again at the end on the Italian day of liberation from fascism in 1945 when the peasants overthrow the feudal power of a farm's *padrone.* The day opens up the possibility of a peasant utopia, but only when seen retrospective of individual moments of revolution earlier in the film: a peasant strike in 1908, resistance of agricultural mechanization after the First World War, the funeral of labor radicals murdered by fascists in 1922. Each past event foreshadows the future emancipation.

Film critic Robert Burgoyne suggests that this is a prototypically Marxist approach to historicity in its recognition that history is not merely a recounting of that which is complete, but is "a particular set of possibilities that can be fulfilled in the future" (Burgoyne 1989: 57). Antonio Gramsci (1971: 82), a major theoretical influence on Bertolucci's film, held that such an approach is at the foundation of history making itself:

> If one applies one's will to the creation of a new equilibrium among the forces which really exist. . . . one still moves on the plane of effective reality. . . . What "ought to be" is therefore concrete; indeed, it is the only realistic and historicist interpretation of reality, it alone is history in the making and philosophy in the making, it alone is politics.

A critical archaeology can create such a vision of the future out of the complex realities it excavates from the ruins of everyday life. These include not only the surpluses, waste, and ruins, but also the remains of futures imagined and lost in the past. Every excavated ruin presages the necessity of founding a new equilibrium in the present.

The households of Lattimer Nos. 1 and 2 had, on a microscale, just such a moment of emancipation when the company divested itself of the town in about 1938. It was a mixed blessing, borne as much of catastrophe as great potential. But unlike Bertolucci's film, which ends on a moment of emancipation that leaves the future open, we are privy to the aftermath, more than half a century that spans this moment until the present.

The chapter begins with the period of decline that led to the collapse of the industry in the 1940s and 1950s. The middle part of the chapter examines the short period at mid-century when the Italian community at Lattimer No. 2, now renamed Pardeesville, elaborated a rich network of communal support outside of the failures of capitalism to provide for their material and social life. This crepuscular era between the generation of the first migrants and their children's children is, even at present, an era of great nostalgia for the community.

DECLINE

Throughout the history of the industry, the financial sustainability of anthracite coal production attained only a precarious economic equilibrium destined to fail, and to affect a great many when this eventuality occurred. Production soared, leading up to its peak during the First World War, boosted by elevated demand and increased colliery efficiency due to various technological improvements to excavation strategies and waste processing (Aldrich 1997: 368–369; Dublin and Licht 2005: 50). Employment in the anthracite industry peaked in 1914 at 180,899 workers (Aurand 1977: 362). In 1917, production of coal by the industry peaked at 100,445, 299 tons (Aurand 1970: 54). Throughout this period, ownership continued to limit workdays to control prices, even during times of highest production. Moreover, fatal accidents in the industry peaked in 1907 with 708 deaths, partially attributable to the expanded workforce and the depth of coal being excavated at the time, but also to new workplace hazards from mechanization such as mangling, electrification, and boiler explosions (Aldrich 1997: 369). Ownership never passed off the benefits of increased productivity to mine laborers.

Following the 1897 strike, solidarity grew among the diverse workforce. In the decades that followed, the coal industry developed the most influential industrial union in American organizing history. Three attributes of the industry contributed to this success: the bonds created among men working independently underground, the mutual support of families in the coal camps, and "the impact of a vast and complex union structure with which those personal loyalties interacted" (Montgomery 1987: 333). The Massacre is significant in this regard as a punctuated moment when the union, and the new immigrant laborers, recognized their interdependence upon each other if striking behavior were to be an effective method of pressuring management (Greene 1964, 1965, 1968).

Decline in the anthracite industry arrived after the First World War with the first drop in demand. Consumer markets turned to other energy sources such as coke, bituminous coal, and petroleum which were cheaper and provided a steadier supply. The consumer market's rejection of the costly product was exacerbated by price fixing by the anthracite combine, which insisted on maintaining a high product price by double dipping on transport costs. Moreover, labor disputes disrupted market supply cyclically between 1902 and 1925.

The finances of the Pardee Brothers Coal Company suffered throughout the 1930s. Records suggest that a variety of factors left the company in near bankruptcy including mismanagement, deteriorating equipment, and infrastructure and, in the opinion of some, the continual drawing of dividends by Pardee family trustees even in the face of failing business prospects (LCC 1935: 17–19). A letter from 1940 shows coal dealers in city markets refusing to sell Lattimer coal due to the failing quality of the product (LCC 1940: 6–24). In a desperate move to avoid total bankruptcy in 1933, the company management was entrusted to the general manager of the collieries, a man by the name of Bob Osler, and a local banker by the name of John H. Wettstine (Foulke and Foulke 1979: 226–238). The dire financial situation prevailed and in 1935 the Pardee Family *in absentia* demanded the colliery operations be liquidated (LCC 1935: 17–19). In letters and memos to employees, management blamed the ownership, deferring threats from labor who suffered greatly from the lack of work. A letter dated to December 24, 1934, and signed simply "the Committee" reads:

> The Men at Lattimer are sick and tired of being idle and half starved. We heard what Aria Pardee told the committee about the coal business is no good and how his people say shut down the breaker and they want half the money. They got lots of money but what the hell are we going to do. No money no work and cant get nothing from the store. . . . We will give you three days to nail a notice on outside your office with names and addresses of these pardee people that say lickerdate Lattimer and we will find a way to fix things. (LCC 1934: notebooks 2–4)

A family history of the Pardee Companies written by descendents completely omits an account of Pardee's decision to liquidate the company in the 1935. Instead it relates the threat to the company posed by the manipulations of Wettstine and others. According to Pardee family historians, in 1936 the family determined that Wettstine had been misusing company funds all along, receiving kickbacks and other payments from employees of local business interests (Foulke and Foulke 1979: 237–238).

Shortly thereafter, company management is turned over to Osler and his associates. By 1938 the company, now officially entitled the Lattimer Coal Company, declared bankruptcy. After a remarkable period of employee ownership described below, the company resumed business briefly during the Second World War during a short period of wartime growth. Ultimately the Lattimer Coal Company shut down in 1945 (Foulke and Foulke 1979: 238).

Interlude: Surplus Enjoyment as the *Place* of Space

It should not be a surprise that the residents of Pardeesville remember their past in ways not immediately associated with its industrial function. Rather, residents reflect upon aspects that arose in excess, or better yet, in *redress,* of this traumatic industrial origin. They remember important unrecorded details about communal interdependence, social intimacy, collective rituals, traditions, and material survival strategies. Much of this history reflects continuities that stretch across the company's divestment of the town into the era of postindustry. What remains of an industrial town when the industry leaves? To answer this question, we must look to the surplus productions of an industrial order.

The concept of the surplus is central to the commodity cycle as the excess value accumulated by capital as it passes through the various parts of an economic system. The paradox of this relationship is that, although the surplus is an *addition* to capital, as the profit motive it is the central feature propelling the process. As this surplus value returns to the capitalist for reinvestment, it alone reflects the continuity of the system. Thereby, surplus value is that value that exists in excess of an object, which then comes to define that object as an externality.

In Lacanian psychoanalysis, as interpreted by Slavoj Žižek, surplus *enjoyment* is a kind of allegory to understand the very nature of all relationships within capitalism including those between people and materials, nonhuman entities, histories, and built environments. Žižek writes that "the dialectics of the commodity-form presents us with a pure—distilled, so to speak—version of a mechanism offering us a key to the theoretical understanding of phenomena." He further proposes that "in the structure of the commodity form it is possible to find the transcendental subject" (1989: 9–10). In essence, *surplus enjoyment* is what is "in us more than ourselves" (Žižek 1992: 200). In such a reading, historical realities emerge only in the dissonance between values: exchange, use, and surplus. And in consonance with psychoanalytic precepts, it is paradoxically only with loss and trauma that the "indispensable condition[s] for autonomy" are conjured: castration, separation, weaning, and so on (Kristeva, quoted in Leach 2005: 183). Enjoyment in Lacanian terms has nothing to do with pleasure: it is in fact *beyond the pleasure principle* (Žižek 1992: 206). Rather, enjoyment is better understood as the negation of pain, a positive condition outside of pain and trauma.

The history of the company town as told by residents and archaeology is a story of surplus enjoyments recollected through trauma. They recall a litany of collective and individual efforts to negate the pain of a landscape of domination and neglect. When the company left the town to its occupants, these enjoyments remained, carrying the town through the subsequent decades. The principles of collective action, anarchism, and radical democracy exist within the realm of everyday interactions (Graeber 2004; Scott 2013). When the structuring agents of central planning are absent, principles of radical democracy can be the natural direction of self-government at a local scale. The short-lived worker's utopia that developed in Pardeesville during this period provides an example, to a certain extent, of unalienated collective action flourishing in the lacunae between divestment and recapitalization.

TOWN LIFE, LATTIMER NO. 2/PARDEESVILLE: 1938–1965

Coal mining company towns originated in the beginning of the Industrial Age when mid–nineteenth-century market prices allowed surplus profit to be reinvested back into the production process. All fixed capital required the monitoring of capital to carefully measure their return investment. This return came in the form of increased productive efficiency or a reliable means to reproduce labor regimes (Harvey 2001). When they no longer serve these purposes, or fail to return investments, capital requires their destruction, recycling, or dormancy. Harvey writes:

> Capitalist development has to negotiate a knife-edge path between preserving the values of past capital investments in the built

> environment and destroying these investments in order to open up fresh room for accumulation. . . . As a consequence we can expect to witness a perpetual struggle in which capitalism builds a physical landscape appropriate to its own condition at a particular moment in time, only to have to destroy it, usually in the course of a crisis, at a subsequent point in time. (Harvey 2001: 247)

In the case of an immobile investment like the company town, this crisis can precipitate a transfer of its role from a machinic system directly responsible for production to a more passive mode of capitalist operation or dormancy for an era of recapitalization to come.

Business records do not directly reference the reasons behind the company's divestment of the towns of Lattimer Nos. 1 and 2 sometime after 1938, but it likely came about as a restructuring demand from debt-holding parties following the bankruptcy of the company. Records, including a collection of maps, memos, and letters, document the complexities of this emancipatory disentanglement. Following company divestment, residents replaced the paternal care of the company with communal decision making and material accommodations already integral to the functioning of the community throughout its industrial life.

The town of Lattimer No. 2 gained an independent identity in 1938 in the form of a new name, Pardeesville, after the family who founded the mines. But from archaeology, digital cartography, oral history, and comparative sources comes a complementary history of rich communal life developing at the time. Collective action in the town materialized both pragmatic material survival as well as that of a more identitarian nature. Of the former, communal infrastructure, scavenging, and food production are reflected in the material landscape. Of the latter, developing communal governance, seizing the means of production, and arranging for the emancipation of the domestic space reflect the development of solidarity along class, kinship, and ethnic lines. In the following sections, each of these tactics are treated separately. In reality, they operated in machinic collaboration, redundantly enunciating forms of collective social consciousness through materiality. In this era, a contravening material dimension also develops as a new regime of spatial differentiation is enunciated among and within individual domestic spaces. Houses grow larger, individual properties expand, and population density shrinks as the intimate spaces of the shanty enclave are dismantled. It is in this period that households in the town begin to reflect contemporary standards of privacy and ownership, a harbinger of what is to follow.

A Brief Worker's Utopia: Employee Ownership in Lattimer, 1938

The spirited revolt that led to the Massacre of 1897 was, to a certain degree, a utopic one. The striking immigrant laborers imagined a better world for themselves, seeking the same kinds of protections offered to earlier generations of nativized American citizens. They also sought reforms that would help all employees and families living under the control of the collieries: the abolishment of the company store and housing system and the enforced use of company doctors. Far from these modest goals, perhaps, was the hope of upending the vertical relationship between employees and employers. Amazingly, for a short period of time in 1938 this did occur. What newspapers described as a "unique experiment" occurred at Lattimer and four other collieries owned by the Pardee Brothers: the workers took control of the company.

In February of 1938 the Lattimer Coal Company went bankrupt, filing papers for a loan with the Reconstruction Finance Corporation (RFC) in Washington, D.C. Immediately, negotiations between the workers, the company, and the United Mine Workers took place regarding the future of the company. In the front of everyone's minds were the nearly 1,000 jobs at stake and the matter of whether the company would pay $65,000 in back wages owed to miners and laborers (*Shamokin New Dispatch* [*SND*], February 17, 1938: 15). Boyd Osler, now in charge, pursued a loan to support a company restructuring effort. In early March the RFC rejected the application suggesting that the efforts of ownership and management were insufficient to save the company.

In March, bypassing the company ownership, employees of the company along with the UMWA appealed this ruling. As a result, the RFC agreed to reconsider the loan (*SND*, March 14, 1938). Negotiations lasted until the end of May with a variety of bids rejected by the committee. By the end of the month, a meeting between the employees of the company and the Hazelton Chamber of Commerce hashed out the beginnings of this "unique experiment" (*SND*, July 18, 1938: 1;

Pittston Gazette, July 18, 1938). By the beginning of June, with company ownership admitting failure, a coalition of officials from the UMWA and the workers presented their own plan to bankruptcy referee William K. Goldstein. In early June, newspapers announced that the 1,000 employees of the company would run the company for a time, using profits to repay themselves the back wages (*Harrisburg Telegraph*, June 1, 1938: 5). On June 1, 1938, the *Harrisburg Telegraph* ran the victorious banner at the top of the paper, "Workmen to Operate Closed Mine Near Hazleton." The plan required that the workers would work for thirty days without pay, and after an additional ninety days would begin repayment of their back wages in 10 percent increments. In a show of support, the Hazleton business community contributed investment money to get the operation under way (*SND*, June 4, 1938: 12). By October $250,000 in wages, including a 10 percent payment on back wages, had been paid and $100,000 in debts had been met (*SND*, October 5, 1938: 7).

Newspapers report that during this period of employee ownership, the company rapidly paid their debts to employees and to the many creditors seeking returns on credit. The experiment was reported in the news as a great success, and miners in similar situations throughout the region studied the operating plans adopted at Lattimer (*SND*, July 26, 1938: 10; December 3, 1938: 2). Representatives from Lattimer reportedly visited nearby mines to consult and share tactics for adoption in similar conditions of company liquidation. Presumably, when the debts to workers were paid, the operations of the colliery were to return to the oversight of the board. Newspapers and company documents provided little to illuminate how this transition transpired. Undoubtedly, the short period of worker ownership was a point of great pride.

An important question remains regarding the negotiations conducted between the RFC, the UMWA, and the workers of the Lattimer Colliery during the early summer of 1938. As the companies divested themselves of the housing in each town, it was passed on to a local real estate company and then sold into private ownership. It is not known if employees of the company negotiated this settlement as part of the restructuring deal.

Hegemonizing Locally: Politics

Around this time, archival documents reveal mounting political struggles arising between company management and residents. A letter written by Wettstine in 1936 to a Mr. Hildun reveals the involvement of the company in manipulating the outcome of local elections, opposing the political successes of a group of unnamed residents of the Lattimer Villages who served as the "ring leaders of the radical element among our employees." Of the companies' perspective, he proposed that:

> [the Lattimer Company is] of the opinion that economies can be affected in the management of the township, Schools, etc., but we also believe that more can be accomplished by co-operation with the present officials than opposing them politically. (LCC 1936: 10–2)

He writes cryptically of their efforts "to strengthen their position as the dominant labor leaders by endeavoring to swing the Township elections to candidates of their choosing" (1936: 10–2)

A New Identity and Spatial Reorientation

Ironically, just as it was emancipated from the family ownership of the Pardee family in 1938, the town of Lattimer No. 2 was renamed Pardeesville. At this time, the town also received its own post office. Previously, residents reported having to walk across the strippings to the post office in Lattimer No. 1. This new identity also included the construction of a road out of Pardeesville with direct access to nearby Route 309 and the removal of the connection between the town and Lattimer No. 1. In the earliest spatial arrangement, between 1869 and about 1908, the only access to Pardeesville was through Lattimer and across a narrow road through the strippings (Figure 6.1). In 1908, the company built a new road extending eastward to accommodate expanded coal workings before curling back to Lattimer just to the east of Canal Street. By 1938 a rough road accessing Route 309 was paved, and the roads between the two towns were removed. At present there is no direct way to travel between the towns.

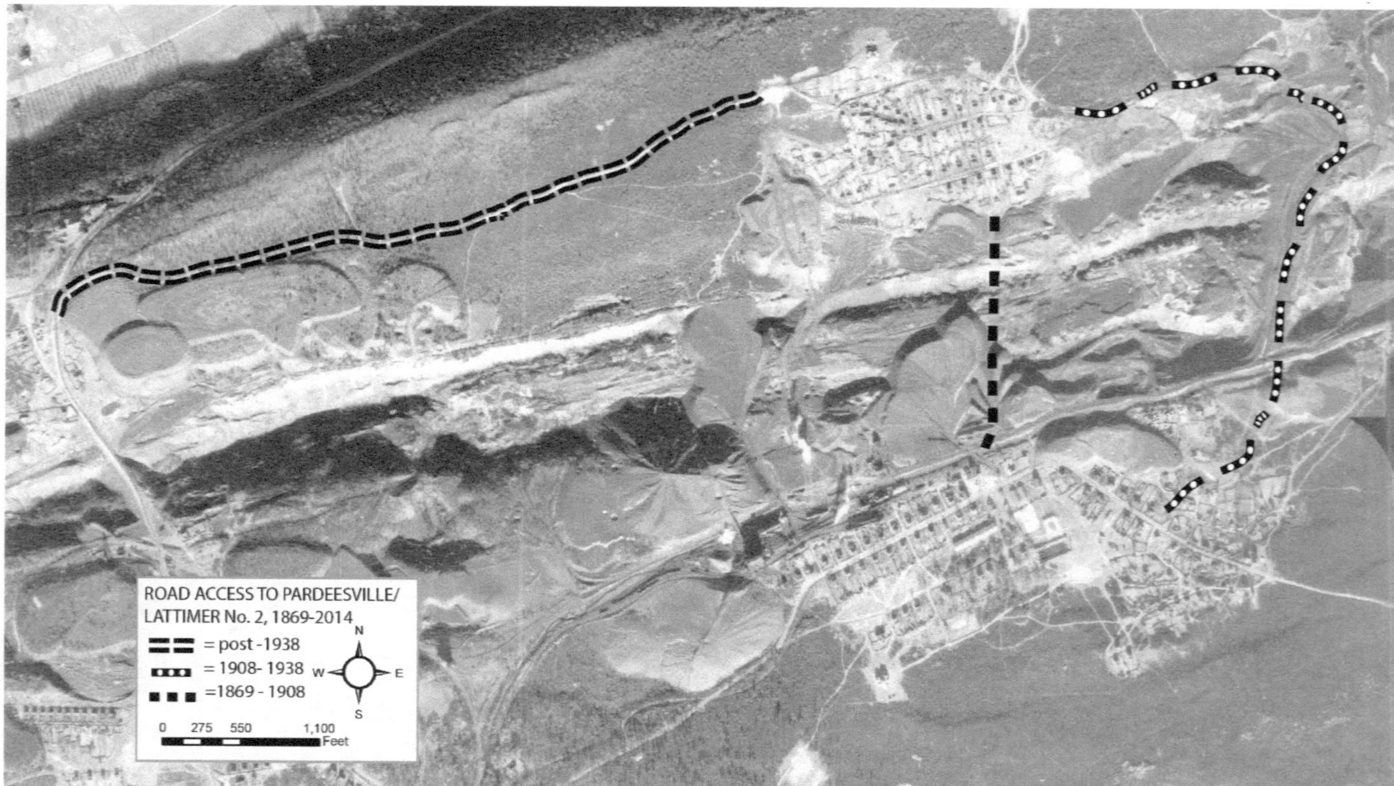

Figure 6.1. Development of access roads to Pardeesville, 1869 to present superimposed over a 1938 aerial of Lattimer Mines (figure assembled by author, base image from aerial photograph 1938, U.S. Department of Agriculture, Agricultural Adjustment Administration).

Communally Organized Infrastructure: Sewage, Water, and Electricity

During the excavation of the Simone Lot in Church Street in Pardeesville, archaeologists uncovered evidence of a communal effort to improve the space of the town following company divestment. A feature identified in Test Unit 6 turned out to be a pipe trench filled with transported soil laden with artifacts dating mostly from the twentieth century. Buried 3.14 feet below surface, the trench was not leveled or lined with sand or gravel as is standard for municipal utility lines from the time period. Made of red terracotta, the pipe measured .6 feet in diameter, large enough to serve as a sewer main. The trench led across the site from the mid-point of the alley, at variance with the known location of the Hazle Township sewers installed along Church Street in about 1985. Another red terracotta ceramic pipe with a width of about .5 feet was identified in a shovel test at a depth of 0.65 feet into ashy soil, indicating it was added after the initial construction of the house. It appeared to lead into the sewer main.

In fact, company records, oral history, and archaeology all reveal that community-organized infrastructure filled in to replace the paternal role the company had assumed for almost a century. Local informants disclosed to us that the pipe is part of a communally installed sewage system connecting house drainage to an unused portion of the mines in about 1943. This informal communal infrastructure is known in the local parlance as a "wildcat sewer." A resident, born in 1931, recalls excavating the pipe trench with the men of the town at the age of 12 (DM, JS, interview, 2013):

> DM: When we put the sewer line in I was about 12 years old . . . I was digging. We were kids. . . . we were digging with a pick and shovel. We had to go to a certain depth, you know. . . . The sewer line . . . went down the hill, and down at the bottom of the hill there was a shaft. And there was a big fan in there and that fan used to blow that air down into the mines. And they ran that sewer line (it was abandoned, all the mines were abandoned). . . . into that big shaft, all the way down to this end.
>
> MR: It wasn't the township that paid for it?

> DM: These people paid for it . . . we were getting a quarter a foot, Mike. But you know what, you were digging. There were two of us together, there was another boy with me, and we were digging to about this depth [gesture to about waist level] to put the sewer down there. We were only this tall, I was about twelve years old, and we had the picks. (DM, interview, November 14, 2013)

A parallel endeavor was also undertaken to provide water to the town. The community excavated trenches and laid pipe to provide water to the town in the 1940s and a committee formed to administer to finances. A letter from the Lattimer Coal Company records revealed that, sometime in April of 1940, a committee from Pardeesville came to the company asking for help in setting up a community water company (LCC 1940: 6–13). The company requested permission to purchase a wooden storage tank located on a nearby coal property for $25. They report that a borehole and well had been dug in town but they lacked the facility to store the water. Written a few months after the initial meeting, the letter reported that the men from Pardeesville were "pressing for help" to acquire the tank and move it to a location on the hill above the town.

The Community of Pardeesville Water Association also secured electrical supply from Pennsylvania Power and Light sometime after 1942 (Figure 6.2). In the meantime, they were serviced by the company. A polite letter thanked the company for use of electricity in the interim. The letter closed with the salutation, "Thanking you for the past favors and assuring you of our fullest co-operation in the future. We are always at your service" (LCC 1942: 17–2).

Community Self-Reliance: Work Diversification

Residents benefited from the efforts of federal or local governance to diversify or subsidize local economy. Diversified work opportunities offered residents some financial relief during and after the collapse of the mining industry. Even during periods of high productivity, low wages of the industry demanded that children find work through legal or illegal means. A 1911 government study of female and child labor determined that children's labor totaled almost 38 percent of total household income (Dublin and Licht 2005: 40).

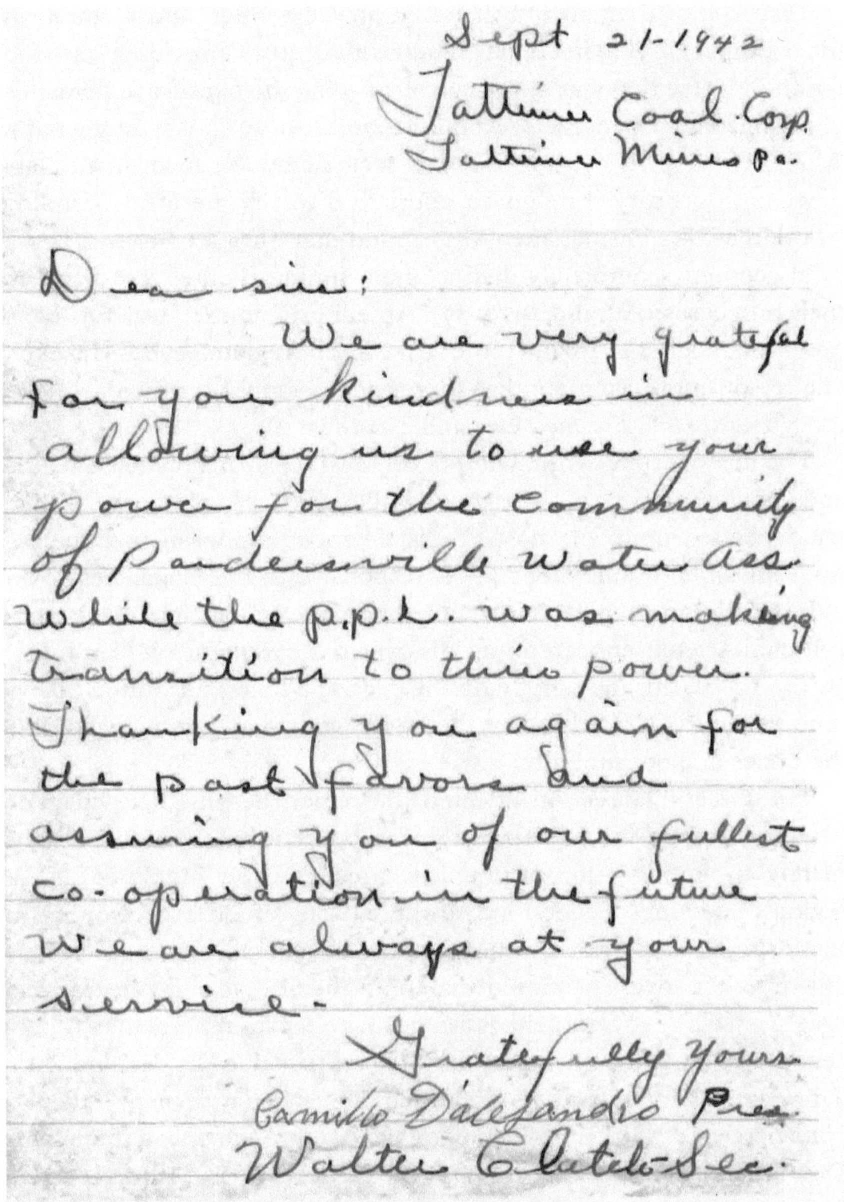

Sept 21-1942

Lattimer Coal Corp.
Lattimer Mines Pa.

Dear sir:

We are very grateful for your kindness in allowing us to use your power for the community of Pardeesville Water Ass. while the P.P.L. was making transition to their power. Thanking you again for the past favors and assuring you of our fullest co-operation in the future We are always at your service-

Gratefully Yours
Camillo D'Alesandro Pres.
Walter Clatch-Sec.

Figure 6.2. A letter from the Community of Pardeesville Water Association to Pennsylvania Power and Light thanking the company for use of electricity in the interim. The letter closes with the salutation, "Thanking you for the past favors and assuring you of our fullest co-operation in the future. We are always at your service" (Lattimer Coal Company Records 1942, 17–2).

Textile and garment industries employed women and children in the region beginning in the late nineteenth century, providing one stop for an industry that was systematically fleeing the high-wage demands of organized labor markets (Aurand 1970, Rose 1981, Westmont 2014, Wolensky 2003). The first mill was secured by the town in the late nineteenth century through an effort to diversify the local economy by middle-class businessmen; they found that during labor unrest the local economy completely died as their mining clientele cut back on their purchases (Aurand 1970: 57). An editorial in the Hazleton *Plain Speaker* lamented in 1887, "If the strike in this region teaches Hazleton one lesson more than another it is that we should have more home industries" (*Plain Speaker*, December 1, 1887: 2).

The post–Second World War period saw the expansion of the textile and apparel industry in the region (Miller 1955). However, this industry depended upon the post-Fordist financial model of maintaining small, flexible mobile labor pools responsive to the exigencies of an industry bound to the accelerated temporalities of fashion. As a result, companies often appeared and disappeared overnight (Miller 1955). By the 1980s, this industry had fully decamped, first for Sunbelt communities in the United States, then overseas, recapitulating each time the cycles of labor mobility.

The Great Depression affected the region heavily. For villagers, however, responses to Depression-era poverty represented perhaps merely an amplification of practices already deeply ingrained in the region's economy. Federal aid to the citizens of Luzerne County in the form of "make-work" programs was substantial. The 1940 census illustrates the diversification of economy and also, to some degree, the lasting effects of government assistance. The census reports that 31 percent of the total documented workforce remained in the coal industry. Thirteen percent of workers labored in the textile or garment industry. Nine percent of workers gained employment through one of three New Deal worker-assistance programs (the N.Y.A., W.P.A., and C.C.C.). Rose (1981: 85) reports that outmigration slowed or altogether stopped during the Depression as "all places appear[ed] to offer the same opportunities (or lack of opportunities)."

In 1947 the business community of Hazleton, in partnership with the chamber of commerce and local media, organized a civic association intended to court new industries to come to Hazleton. After a few false starts, the CAN DO Corporation was formed in 1956 to fund infrastructure development to encourage new industries (CAN DO 1974; Rose 1981: 110–115). At the time, the chamber of commerce offered concession fees and tax breaks to businesses willing to locate to the area. A community drive urged every citizen to contribute a dime a week for a full year. In time, the organization raised enough money to purchase land for a business park. Other projects followed, bringing a variety of light and heavy industry to the region that provided employment to the generation growing up in the 1940s. For them, the coal industry was an economic structure mostly confined to the world of their parents and grandparents, although the landscape and heritage inescapably surrounded them. CAN DO was largely a success in turning around the economic atmosphere of the 1950s. Perhaps its biggest success was in securing the alignment of Interstate Highways I-81 and I-80 to pass just on the edge of town. CAN DO is still active in the region, but as models of flexible accumulation took hold of American industry in the 1970s, Fordist models of industrial community could no longer compete in a globalized market.

There is some indication that the workforce in the region retained strong tendencies toward labor resistance, but the labor history of the period after the 1940s is yet to be written. In 1972, just as the power of collectivized labor was waning in the United States, union leaders from all over the country met with local unions to commemorate a monument to the Lattimer Massacre near the site of the tragic event. The granite monument was erected by the United Labor Council of Lower Luzerne and Carbon Counties along with the UMWA and the AFL-CIO. Scholars, activists, politicians, and working men and women were all present at the event. Cesar Chavez, historians Edward Pinkowski and Harold Aurand, and popular Pennsylvania congressman Dan Flood were all present. The Lattimer Band, composed mostly of Pardeesville residents and directed by Reverend Ferrara, played the national anthem. In a speech given at the time of the commemoration, George Meany, president of the AFL-CIO at the time, stated, "The things [the fallen miners] died for have not yet been achieved" (*AFL-CIO News*, September 30, 1972: 3).

Communal Self Reliance: Gardening, Food-Sharing, and Scavenging

Interviews with residents of coal patch towns in the region report a variety of practices that contributed to the economic self-reliance of communities in the past. Residents of the ethnic enclaves of Italian dwellings in Lattimer and Pardeesville particularly connected these experiences to shared experiences of poverty and communal solidarity. A resident of Pardeesville recalled:

> They all had to work. Who worked in the mines? Who worked down in the farm? . . . If you look close, that's all. So, one wasn't any better than the other. They were all in the same pot with boiling water. . . . getting boiled slowly. (MD, interview, March 21, 2013)

The domestic landscapes of the patch towns always included backyard gardens planted "property line to property line" (AM, interview, November 14, 2014). Features found in the base of test units throughout backyards attest to the varied activities of planting gardens, including posthole setting, shovel scrapes, and other shallow depressions from the planting, tilling, and the supporting and fencing of plants.

In interviews conducted in the region, residents suggest that this tradition of home vegetable growing helped families stretch limited resources and scant wages. It also lessened a family's dependence on the company store. A former resident from the nearby patch town of Eckley described his experience in a Slovak household (JM, interview, March 20, 2011):

> JM: Oh everybody planted large yards ya know . . . maybe they were like thirty feet wide and a hundred feet long. And they canned, they canned all the vegetables and stuff. And a quick story, my parents, like most people, would buy cabbage in fall and make sauerkraut. They would shred the cabbage, they would put it in the barrel, and here's how they stamped it, Mike. My father would wash his feet, he'd put on his Sunday "go to church clothes," and he'd jump in the barrel to crush the cabbage with his feet . . . his bare feet. Then we would eat it later. So you would make the sauerkraut, and then mostly everybody had a cow or

> pigs or goats. . . . And I remember Mom a couple times, we'd go shopping and she's counting and she'd have five dollars maybe, and she's counting. Ok now this is nineteen cents, this is eighteen cents . . . I'm up to a dollar twenty, I could do this and do this. . . .
> MR: Was there a company store near you?
> JM: There was a company store in Eckley. Now my father made it a policy not to buy anything there except gasoline for his car and kerosene, for the kerosene lamps. 'Cause their prices were out of range, ya know. That was his policy.

By all accounts, gardens served as a kind of bank to survive unemployment caused by all manner of reasons: economic fluctuations, natural disasters, and strikes.

During lean times, the securing of food often became a collective project (Miller and Sharpless 1998: 314). In the socially intimate settings of "the Hill" in Lattimer and "the Village" on Church Street in Pardeesville, the sharing of food between residents served to tie together the bonds of solidarity. A former resident of "the Hill" in Lattimer describes growing up in this neighborhood in the 1940s and 1950s (AM, interview, November 14, 2014):

> "The Hill" was like a commune. "The Hill" was self-sustaining. Everybody had animals. Nobody cut grass, they planted it from property line to property line. There was big wooden fences. When I was a little boy those fences looked real big. . . . We had rabbits. We always had like eighty rabbits. Cerese had chickens, Bagnatas had goats, and when the Gerhardts moved and [the] Yanacs . . . he had a farm up here . . . he had pigs over here, he had chickens and ducks.

Talk of food turns naturally to the topic of communal upbringing and sociability:

> When we were young, somebody was always related to somebody somehow. I don't know how, but there was a lot of relations in the area here. We would visit each other and it was like family that visited family. Nobody visits like that anymore. Nobody even knows their neighbors anymore. Before I used to eat on every house on "the Hill," every day.

Food was traded between families, with each family specializing in something.

Scavenging and banking architectural materials, fuel, and other supplies from dumps provided families with materials for construction, renovation, and repair. Architectural items recovered from deposits at Canal and Church Street suggested materials came from a great variety of sources. Scraps of items such as sheet metal, leather, rubber, and brass with cut marks and alterations also suggested use as patches for repair. A resident of Pardeesville who still maintains a large collection of assorted items described the tendency of residents to collect as a form of survival (MD, interview, March 21, 2013):

> Well I mentioned that people were poor. Why throw it away because tomorrow, you may need it? My family, we were very poor, and so were a lot of other people. So whatever we had, well just put it on the side for the time being, stick it in the shanty, or down in the basement. And people like me, which you saw earlier in my cellar. . . . and this is only a part that you are seeing there, there is more. . . . Why do you think people hoard that stuff? Because there is a fear. A fear develops in a person that, 'well what if tomorrow comes or somewhere down in the future I need this item, I don't have it. What am I gonna do? But if you store it away somewhere . . . Yeah, yeah . . . I remember having, somewhere, something, a couple of boards or whatever . . . in the garage or in the shanty. Maybe I could use them. Sure. I remember as a kid going around these dumps; looking for old bicycles. We used to patch the tire with tar. 'Cause things weren't around, whatever we could. Go around the dumps, see if we could find an old tire.

Picking coal from refuse banks was reported by nearly all interviewees as a major contribution by women and children to household economy.

Seizing the Means of Production: Bootlegging Coal

During the period of decline, subversive strategies for seizing the means of production were also adopted or intensified. An intensification of coal picking can be seen in the practice of bootlegging, prevalent in the anthracite region during the Great Depression (Dublin and Licht 2005: 76–82; Miller and Sharpless 1998: 319). Bootleg or independent operations were composed of groups of three to five miners, often families or close associates, excavating coal from outcrops, banks, and even excavating small shafts and slopes. Coal excavated in this way by independent miners illegally on public or private land throughout the anthracite region made up for low or nonexistent wages during times of unemployment. It was often picked up by a network of independent truckers who drove the product directly to city markets, sometimes selling it house to house. In the process, they bypassed the regulating hand of the railroad monopoly (Dublin and Licht 2005: 76–82). Writing about workers' responses to unemployment for *Living Marxism*, Paul Mattick generalized about the practice of bootleg coal mining (1938: 90–91):

> The problems of all workers are here, so to speak presented in a nutshell. All that is really necessary for the workers to do in order to end their miseries is to perform such simple things as to take where there is, without regard to established property principles or social philosophies, and to start to produce for themselves. . . . The bootleg miners have shown in a rather clear and impressive way, that so much bewailed absence of a socialist ideology on the part of the workers, really does not prevent workers from acting quite anticapitalistically, quite in accordance with their own needs. Breaking through the confines of private property in order to live up to their own necessities, the miner's action is, at the same time a manifestation of the most important part of class consciousness,—namely, that the problems of the workers can be solved only by themselves.

Bootleg coal made up 5 percent of the coal brought to market in 1936–37, expanding to almost 10 percent by 1940 (Dublin and Licht 2005: 81, Mattick 1938: 86). There is evidence for bootlegging on land owned by the Pardee Brothers in company records. A letter from the Coxe Brothers Coal Company dated July 16, 1936, brings attention to "considerable bootlegging" that was "being done in the Old Mammoth Stripping east of Drifton #9." The letter suggests that under the terms of their lease, the Pardee Brothers are responsible for seeing that "the property is protected," and the company, "must take whatever steps are necessary to see that this bootlegging is stopped at once" (LCC 1936: 14–33). In another letter between the general manager of the Lattimer

Coal Company and the general manager of the company store dated April 27, 1936, he suspects bootlegging or coal picking as the cause of several local individuals having "not bought any coal for some time." The writer directs the manager to "please speak to them in that connection" (LCC 1936: 14–36).

Because of the sympathies of the public, law enforcement, and many politicians for the plight of miners in the region, industry was unable to curb the practice of bootleg mining until 1941. Around this time, a semipublic organization known as the Anthracite Emergency Committee, made up of operators, union officials, and public representatives, was created to set rules requiring coal operators to meet production quotas, hire unemployed miners, and purchase coal from independent or bootleg operations. In effect, bootleg operations were absorbed into the preexisting capitalist structure. When the Second World War briefly increased demand for coal once again, bootlegging declined as miners were again employed under the traditional structures of the operating economy (Dublin and Licht 2005: 81–82).

Social Banditry: Forms of Southern Italian Criminal Resistance

Historic newspapers from Hazleton from the early part of the twentieth century included many stories of bandits haunting the hinterlands around town. In these accounts, bandits were connected to the notorious organization known as the "Black Hand," as the Mafia and the Camorra were called. Interestingly, the stories often implied that friends, family, and community in the shanty enclaves of Lattimer aided some criminals, hiding them from authorities. For example, on July 9, 1909, the newspaper reported that Frank Mauro, wanted for the murder of Michael Mucillo, was hiding in a swamp at Lattimer. He was surrounded by a posse, which had been granted permission by the law to "shoot him dead" if he attempted to escape. The newspaper reported that law enforcement suspects that friends and relatives had been bringing him food and water. On the 12th of July, 1909, Mauro escaped and left for Italy, his passage paid for by relatives. On January 27, 1910, masked highwaymen held up taxicabs on Harleigh Hill Road north of Lattimer. Armed with revolvers, they used logs to block the road. On May 22, 1913, Joseph Abramie fled into the nearby mountains after stabbing Sandy Malavento. Police searched the Italian enclaves of several towns including Lattimer. Evidence gathered by police suggested Abramie was an agent of the Black Hand, connected to a band of cutthroats, and Malavento was killed for fear he would "divulge facts to the police concerning their operations."

Eric Hobsbawm has outlined forms of preindustrial collective resistance in his book *Primitive Rebels* (1959), citing them as the precursors of organized socialist or political rebellions. Indeed, Southern Italy has a long social history of banditry throughout the nineteenth and early twentieth centuries. Many of these individuals achieved a sort of local fame as Robin Hood–like antiheroes, and hagiographic accounts of their magical capacities to avoid capture are detailed. In fact, the social banditry of the time cannot be separated from the close bonds of kinship which, when confronted with a destabilizing force from outside, throws it into disruption. The political ambitions of banditry, Hobsbawm contends, are modest: "a traditional world in which men are justly dealt with, not a new and perfect world" (1959: 5). But as this confrontation deepens, and communal and kin groups become differentiated by relative wealth, Hobsbawm suggests that class struggle may enter into consideration as the traditional modes of social leveling, including blood vengeance, may become broadly socialized. Such a change is documented in parts of Italy during this period.

The Religious Community in Pardeesville

Between the 1930s and the 1970s, Pardeesville was mentioned in the newspaper most often in either the death notices of the first generation of immigrants or the wedding notices of their descendents. Following these were a vast number of announcements connected to the St. Nazarius church festivals or the meetings of various religious social groups. The early landscape of the shanty enclave epitomized the regionalism of Southern Italy known as *campanilismo,* the literal translation of which is "the sound of the bell." The term identifies a rural practice of small town residents whose identity was tied to the space within earshot of the church bell (Cohen 1990: 87; Williams 1969 [1938]: 9, 34). The earliest maps of the town reveal that residents built a church at the time the first residences were constructed in the northwest corner of Lattimer No. 2. Local lore recalls that the first masses were celebrated in 1879 in a tent made from resident's sheets (*Plain Speaker*, July 31,

1954:3). In 1884 residents constructed a wooden church built of collected scrap wood. It served as a mission church with masses delivered by a travelling priest from Hazleton. In 1904 the church was expanded and a stationary pastor established in town. An elder resident recalled the communal construction of the old church:

> this church was made by the town's people. Wherever they got the lumber, wherever they got the wood. There were, in those days I remember the "dooly" boxes, they called them "dooly." . . . dynamite. When they took the dynamite out of them the men took them and they used it for sheeting on the church. And [it was] made from all different kinds of wood. They built that themselves. (DM, interview, November 14, 2013)

The most visible events were the yearly festivals that drew thousands of visitors from out-of-state to partake in entertainment, food and to observe the traditional Italian practice of processing the statue of the saint throughout the town. The practice has its origin in Southern Italy, and transferred to the states:

> The feast-day celebrations of the patron saints, therefore, are conducted in much the same fashion as they were in Italy, with church services followed by processions through the streets of the district. The devotees still make offerings during the march, ask or recognize favors in the traditional way, and then end the affair with fireworks, music, singing, and feasting. (Williams 1969 [1938]: 149)

Residents celebrated the feast day in a similar way in Pardeesville throughout the century. The last celebration was in 2009 when the parish was closed by the archdiocese. A resident of Pardeesville, born in 1943, recalls the communal effort required to put on these events (MD, interview, March 21, 2013):

> I remember we used to have these festivals here. People got together. They worked hard to prepare for them. And it was a happy thing. We had the band and the procession in honor of the saint. Wherever it was, wherever the community you were in. And it was a wonderful time. . . . All this stuff started over there in the old country and it exists today.

Figure 6.3. Procession of St. Nazarius in Pardeesville, 1930s (photo printed permission of Grace DeLorenzo).

To the right and left of the statue of St. Nazarius were *cinte,* or ornate constructions adorned with candles carried by community members. Strings of dollars bills tied to the statue are visible, an American adaptation of the Italian tradition of offering fruit to the saint (Cohen 1990: 92).

In 1947 residents constructed a new church on the hill above town, moving the social center outside the tight confines of the shanty enclave. This paralleled the movement of residences away from the restricted space of the enclave [as discussed in Chapter 7]. As the local economy improved in the decades following the successes of CAN DO, the St. Nazarius festival grew in visibility and scope. In the 1960s and 1970s it expanded from a one-day celebration to a three-day festival featuring a talent show, prizes, and a bazaar. A variety of mutual aid and religious societies conducted the planning, financing, and support for these events. Among these groups that grew around the church were organizations such as the Blessed Virgin Mary Sodality, the Assunta

Society of Pardeesville, and the St. Nazarius Holy Name Society. The traditional procession around the town, however, was its primary attraction.

Lizbeth Cohen asserts that the religious procession amounts to a ritual uniting secular and religious identity. The procession brings the saint "out of the church and into people's everyday world. As villagers paraded behind the saint and joined in the celebrations afterward, they not only sought penance or favor, but also declared loyalty to this unified religious and secular community" (Cohen 1990: 88). Viewed through a traditional ethnographic lens, festivals are interpreted as opening up liminal zones of inversions. In contrast, the feasts of Italian Americans beginning in the early- to mid-century "clarified the way the community was evolving" (1990: 92). In an American context, the implications of the ritual changed, particularly in the context of urban settings in which neighborhoods were far less spatially bounded than the villages. Cohen contends that Italians "used the symbolic power of their traditional feasts to express their neighborhood's rival claim on religious ritual and independence from the church hierarchy" (Cohen 1990: 89). As communities grew increasingly distant from the bounded spatiality of Old World neighborhood or village, the feast days were reoriented as public expressions of festivity open to a broader Italian-American identity. At the same time, a focus on monetary donations reflected a new emphasis on the materialistic, capitalistic, and political dimensions of the proceedings (Cohen 1990: 92). The political and economic visibility of the festival and the integration of its organizing apparatuses into the fabric of a postcoal world paralleled the role many of the town's residents would take in shaping the local politics of the region in coming decades.

CONCLUSION

Historicity is a semiotic operation whose limits are the limits of semiotics itself. For Lacan, the "Real" coincides with the surplus enjoyment that erupts in excess of symbolization:

> The Real is at the same time the product, remainder, leftover, scraps of this process of symbolization, the remnants, the excess

> which escapes symbolization and is as such produced by symbolization itself. (Žižek 1989: 191)

Archaeology optimizes such an approach to the historicization of the Real, with its focus on undocumented waste, ruins, scraps, and surpluses.

Places are assemblages, writes postcolonial theorist Achille Mbembe, "enclose[ing] multiple durées made up of discontinuities, reversals, inertias, and swings that overlay one another" (2001: 8). Materiality such as the built environment maintains its own temporalities and memories that often endure beyond the intentions of their creators (Hodder 2012: 100; Olsen 2010). Individuals and communities remain entangled with commodities, landscapes, and the habits associated with their use long after their exit from the commodity cycle (Dawdy 2010). Dialectically binding the *longue durée* of materialist history with a nuanced seriated approach to community history reveals the dialectical tensions of continuity and rupture that create a place.

We can simply define the Real as that which "always returns to the same place" when all is said and done (Žižek 1989: 50). By mid-century, King Coal had abandoned the region, leaving behind a landscape in tatters. When these absentee owners picked up and left, they "abandoned their collieries and leased or sold their lands to the industry's scavengers: surface strippers and small time operators" (Miller and Sharpless 1998: 321). The residue of industry, the company towns and their residents, remained thick on the landscape, armed with the materiality of survival from decades of struggle: the weapons of the weak. Emptied of the productive mandate of its industrial machinery, the town suddenly embodied nothing other than its surplus value: enjoyment *or at least the means to freedom from pain.* For a time it continued to be what it had always been.

The community survived a potential catastrophe in the 1940s when the company divested itself of the town. Changing immeasurably in this era, the residents charted a movement akin to a massive migration: of identity, of function, and of power. As mentioned earlier in the chapter, J. H. Wettstine's letter to a fellow coal company operator avowed the company's opposition to the political aspirations of the radicals of the village. He endeavored to upset their efforts to "strengthen their position as the dominant labor leaders." In this assessment, Wettstine missed the direction of local history. Viewing power in the region from an anachronistic industrialist perspective, he could only perceive of the opposing forces in the region being those of capital and labor. In truth, the village radicals' political aspirations went much further. If it was not clear before, by mid-century the Italian community on the periphery of the former company town had claimed this space as coextensive with the very center of their community. In the yearly procession, their patron saint had already circumambulated the company-owned space for perhaps at least half a century, dollar bills pinned to his robe, marking it as its spiritual and political kingdom.

The danger of fixating upon surplus is to risk denying a critical attentiveness to the epochal cycles of capitalism, the eventual *return of the repressed*. It would be a mistake to assume that the exit of the coal industry meant the withdrawal of capitalist logic from the spatiality of the town. In fact, the ferment of a brief hiatus is often the very foundation necessary to "reestablish the basis for renewed accumulation" (Harvey 1985: 116). As argued by Žižek (1989: 53–54), "the 'normal' state of capitalism is the permanent revolutionizing of its own conditions of existence: from the very beginning capitalism 'putrefies.'" As the normal state of capitalism is that of crisis, the withdrawal of the industrial regime paved the way for new forms of recapitalization. These changes came along with the dawning of a new national economic order variously described as flexible accumulation, post-Fordism, or neoliberalism, described in the next chapter (Harvey 1990, 2005; Read 2009; Rose and Miller 1992).

7

The Destructive Character

Renewal and Memory

> The destructive character knows only one watchword: make room. And only one activity: clearing away. His need for fresh air and open space is stronger than any hatred. . . .
>
> The destructive character sees no image hovering before him. He has few needs, and the least of them is to know what will replace what has been destroyed. . . .
>
> Because he sees ways everywhere, he always stands at a crossroads. No moment can know what the next will bring. What exists he reduces to rubble—not for the sake of rubble, but for that of the way leading through it.
>
> *Walter Benjamin (1978: 301–303)*

This fragment from Walter Benjamin deals with the contradictory character of progress and destruction in the twentieth century. The destructive character makes ruinous, but with its "need for fresh air and open space," it is also productive. Destruction "make[s] room" for progress. It has a revolutionary character, but it has no clear idea of what comes after.

Benjamin's fragment perfectly describes the direction urban renewal took in the city of Hazleton beginning in the late 1960s. In the name of "fresh air" and "open space," almost 200 buildings were demolished with little put up in their stead. A somewhat more restrained destruction took place in the township surrounding the city. Patch towns such as Lattimer and Pardeesville were recapitalized into suburban spaces, while shopping strips drew commerce, energy, and focus away from the urban core.

This chapter begins with archaeological evidence for the spatial atomization of the shanty community at Pardeesville beginning in the second half of the twentieth century. The renewal that followed encompasses the period of destruction preceding the recapitalization of the town's space. Two planning documents from the 1960s chart the broad economic and political direction the landscape of the town took in consonance with the broader ideology of the Housing Act of 1954. These telling documents show how a new financial logic, vertically integrated on a national scale, penetrated into the most intimate of spaces. This ideological turn was mapped onto the material life of the residents of Pardeesville in the years after the divestment. These changes obliterated aspects of the material, social, and political landscapes that provided working class people of the region degrees of pleasure, solidarity, and power in previous industrial eras. Their destruction and replacement by juridical structures, governmental identities, and negative freedoms produced new ideologies and subjectivities leading to alienated communities of autonomous individuals networked to regional governance. Arguably, these conditions produce aspects of symbolic and structural violence far more brutal than the coercive capital-labor tensions of previous generations, albeit dispersed and largely indiscernible (Žižek 2008). As Benjamin warns, the "destructive character" operates in many forms, "not always by brute force" (Benjamin 1978: 303).

THE ARCHAEOLOGY OF COMMUNITY CHANGE

The razing of the residence at Church Street in Pardeesville sometime in the late 1960s or 1970s is as much an instance of historical rupture as continuity. More than simply the terminal point of archaeological history, the pattern of razing and movement throughout the village at the time is one moment along the trajectory toward community atomization. These changes are visible throughout the archeological record, materialized in spatial alterations to houses inflected in differentiation, expansion, reconstruction, razing, and patterned migration. Moreover, the patterned destruction of the housing along Church Street, far from mere postdepositional or taphonomic factors, has made archaeological research possible here by providing an empty, unoccupied, open grassy lot at the center of the former shanty enclave. The archaeological process is, of course, the most final act of physical destruction.

The first structure built along Church Street in Pardeesville was a long barracks-like building present by at least 1885. In contrast to the rigidly enforced homogeneity of the company double houses, shanty occupants were free to alter site plans in response to changing needs and sensibilities. The first half of the twentieth century reflects cumulative additions to the structure made by residents to accommodate boarders and expanding families. As discussed in Chapter 4, these changes also indicated the differentiation of space in response to changing settlement patterns and sensibilities of space and privacy. A map from the 1940s reveals a literal rupture in the pattern of cumulative growth in the form of a narrow (≈3.0 ft.) alleyway separating what appear to be two separate households. Test units investigating the narrow passageway between buildings, however, failed to turn up any subsurface indications of the passage, although its presence was recalled by former residents of the house. Instead, a 2-foot-thick stone foundation (Feature 8) was uncovered along the eastern wall of the residence. To the west, archaeologists identified the remnants of the 1885 brick foundation, interrupted at less than a perpendicular angle conforming with the outline of the brick barracks. As discussed in Chapter 4, where the alleyway should have been, we uncovered a concrete cellar packed with architectural and domestic debris ranging in date from the turn of the century to the 1960s.

Company maps dating to the 1940s reveal that a real estate company began to selectively demolish houses in the shantytown when they could not be sold to renters, presumably to reduce housing density in the enclave. This patterned demolition reflected an effort to alter the density and aesthetics of the town's built environment in conformity to modern norms of privacy and space. A real estate map dating to the 1940s shows the easternmost unit of the structure marked "to be razed" (Figure 7.1). Likewise, the pattern of selective removal is pursued along all the domiciles fronting Church Street. The cellar represented yet another phase of the building's renovation which saw the consolidation of the once thrice-divided barracks into a large single-family home. Residents accomplished this by demolishing and replacing the two easternmost units with a cellared extension.

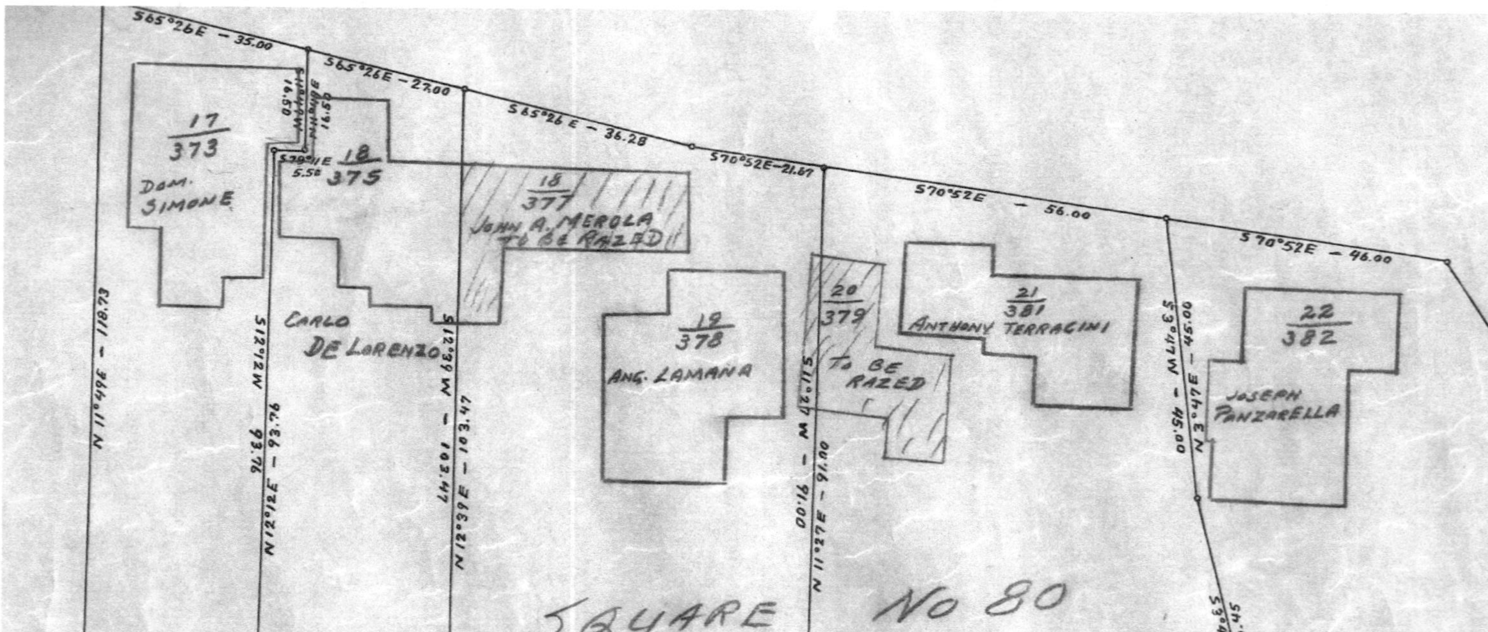

Figure 7.1. Real estate map of Church Street in Pardeesville from the 1940s (Lattimer Coal Company Records 1940s).

Figure 7.2. Plans of the domiciles around Church Street, redrawn from various real estate and company maps and aerial photography showing decreasing population density and larger house footprints (figure by author).

Architectural features examined in combination with company and real estate maps and aerial photography from between 1878 and the present reveal radically shifting spatial patterning in the shanty enclave in Pardeesville. The resulting changes reflected a movement over the course of about sixty years of increasingly atomized spaces and expanding house sizes within the densely populated boundaries of the shanty town. House sizes expanded considerably after 1940 as did the spaces between them (Figure 7.2).

The most radical changes to the town landscape arrived after the Second World War when renters were given the chance to purchase new lots developed by Hazle Realty Company along a new access road into town. Many residents who chose not to purchase lots in town purchased along the new byway. Many members of the younger generation also moved up the hill to start their own families. An aerial photograph from 1959 depicts about thirteen such dwellings. By 1969 about twenty-five are visible. Family migrations, reconstructed from census info and real estate maps dating to the 1960s reveals the movement of household members at the time, generally from the western portion of town (Figure 7.3).

Local residents reported at this time that garden soils were often excavated and replanted around houses on the larger lots on the hill. A resident of the town reported:

> I have some [soil] here from my mother's. They lived down by the strippings. There was an old house there. They took the topsoil and I got the one below that. A lot of them, they took the topsoil wherever they could get it. From the old houses they dismantled and nobody bought the property and the topsoil was there and they took it. (DM, interview, November 13, 2013)

In some cases shanties were deconstructed and their materials stored or integrated into outbuildings or other structures. In comparison to the aging houses in the shanty enclave and elsewhere in the town, the houses on the hill resembled typical low-density suburban housing of the mid-century construction.

The landscapes of Pardeesville that evolved throughout the second half of the twentieth century did not merely reflect the choices of residents alone. If residents' motivations reflected a continuation of previous efforts to better their lives, capital and government channeled these labors into ways to recapitalize the ruined industrial landscape for profit, taxation, and a revitalization of the labor market. Throughout these developments a remarkable consonance, or convergence, was struck between actors at various scales of government, business, and community, each of whom operated within the principles of entrepreneurial spirit, fulfilling recapitalization as financial logic penetrated into everyday realities.

Figure 7.3. Graphic depiction of family movement from company town spaces to new development along the new Pardeesville access road, reconstructed from census and real estate maps (figure by Adrienne Allegretti and author).

URBAN RENEWAL

In July of 1973, a published photo of the south side of the city showed tall weeds growing on the empty lots where buildings once stood (Figure 7.4). Beginning in 1968, the city of Hazleton used grants provided by the federal government through the Housing Act of 1954 to establish a redevelopment authority. They endeavored to remake the failing postindustrial city. Demolition focused upon the area known as "Downtown South," approximately sixteen square city blocks. Of a total of 225 buildings in the area, the redevelopment authority demolished 198; 122 were residential structures. The city used eminent domain to seize properties, providing minimal reimbursement to landowners. The Hazleton Redevelopment Authority was taken to court time and again for reimbursing owners a mere fraction of the land's purported worth (Tarone 2004: 129–132). When the dust settled, many historic buildings were demolished with little left in their place (for "the Destructive Character . . . has few needs, and the least of them is to know what will replace what has been destroyed." [Benjamin 1978: 301–303]).

In time reconstruction commenced in Hazleton, but the destruction of the city that continued into the early 1980s remains a traumatic issue for elder residents. Joe Michel, a project collaborator and former city engineer, maintained a collection of bricks as a memorial to each of the prominent buildings whose destruction he witnessed. These included many theaters, hotels, and the Hazleton train station. An in-depth accounting of the motivation and organization behind the destruction of the town has yet to be conducted (Tarone 2004: 131).

Urban renewal originated as a solution to two major problems the nation faced after the Second World War. Maintaining the competitiveness and steady expansion of Fordist industry in the postwar period demanded certain levels of infrastructural efficiency and dependability. Moreover, even with the economy in full swing, capital remained poorly distributed among sectors of the workforce and the unemployed. The state, coffers flush from rapid expansions in tax flow, was called upon to provide for the redistribution of capital on behalf of the national economy through investments in infrastructure, transportation, social wages, and housing (Harvey 1990: 133–135). Urban redevelopment was one such deployment. Initially, its focus was the reshaping of American

Figure 7.4. Image posted in the *Hazleton Standard Speaker* on May 5, 1973, showing weeds growing in the empty lots of downtown Hazleton after they were left vacant by urban renewal. The newspaper points out the irony of the city's refusal to obey its own ordinances against weeds (photo from the *Hazleton Standard Speaker*, May 5, 1973).

cities through "slum clearance" and other efforts to transform residential city life. The Housing Act of 1949 stimulated the spread of redevelopment to many cities by providing federal funding for the acquisition and demolition of buildings declared as slums and for low-income housing to provide for the displaced (Self 2003: 142–143; Fairbanks 2012, 2014).

The passing of the Housing Act of 1954, however, provided a new emphasis on economic renewal. Providing for greater collaboration between government and business, the act essentially allowed federal funds to finance local efforts to do practically anything they wanted in the way of redevelopment. In essence, what began as an effort to provide low-income housing to the most vulnerable became "a program to rebuild downtown" led by the business community (Self 2003: 143; Fairbanks 2014: 5). The directives of the 1954 act guided local redevelopment authorities to target the broader social contexts of "blight." As a comprehensive plan, it sought to administer to what it considered sustainable improvements in social and economic growth as opposed to the "piecemeal thrusts" of earlier generations of urban redevelopment (Fairbanks 2012: 1). A document prepared by an advisory council to the president, comprised of business leaders and social progressives, suggests that rather than simply demolishing slums, renewal should be aimed broadly, encouraging "a standard of decency and order which will support the self respect of people, adequately serve family life, and prevent physical surroundings which are unsafe or unhealthful or destroy community and citizen morale" (U.S. Presidents Advisory Committee 1953: 134). A 1961 amendment increased the share of federal funding offered to municipalities with fewer than 50,000 people. As a result, small municipalities such as Hazleton and surrounding Hazle Township were given incentives to take part in the national program (Fisfis and Greenberg 1962: 61).

Hazle Township's redevelopment efforts began in earnest in the late 1960s. *The Hazle Township Comprehensive Plan* (Kendree and Shepherd 1967a, 1967b, 1967b) was funded through an urban planning grant from the Department of Housing and Urban Development under the provisions of Section 701 of the Housing Act of 1954. It was prepared by Kendree and Shepherd Planning Consultants of Philadelphia. It is divided into three sections, a Development Conditions Study (1967a), Development Proposal (1967b), and a Comprehensive Plan (1967c).

Read "along the archival grain" (Stoler 2009), documents of "unrealized plans . . . short-lived experiments, and . . . failed projects" provide a rich ethnographic source to view the motivations and ideologies of political economy. Buried beneath detailed demographic and economic statistics of the township is a detailed plan for the recapitalization of the region's ruined material, social, and economic life. The built environment of the coal company towns and other traces of industrial origin—such as unsightly mine waste, underutilized open space, population nucleation, and exaggerated stratification in salaries, housing, and commercial development—are cited as "the greatest drawback to industrial development" in the region (Kendree and Shepherd 1967a: 5-2–5-4; 1967b: 1-6). Moreover, many of the materialities of community resistance and support described in Chapter 6 are coded and targeted for elimination in the report: banked salvaged, gardens, informal and communal spaces for socializing, community economy, and spatial intimacy.

Particularities of each patch town are addressed, evaluating redevelopment potential as well as barriers. Pardeesville is described accordingly:

> There is no industry in the village except for a small coal stripping operation in the eastern section. Pardeesville has a small post office which also includes living quarters and a small store. Food purchases can be made from a converted bus which visits Pardeesville from Lattimer. Public transportation is available. The village is made up primarily of single-family detached houses with a few twin houses. Approximately 20 percent of the housing is in need of major maintenance. The biggest problem facing the community is the storage of trash and junk within the village. (Kendree and Shepherd 1967a: 3-1)

Elsewhere the report evaluates development potential around the town. It cites the attraction of the natural setting, which is, ironically, upset only by the blighted village itself:

> The most significant development has occurred in the western section along the Pardeesville Road. Here there are about 15 new homes. . . . The area to the east and west is composed of small pine trees primarily. It has not been stripped. By and large the area has an attractive location, with the exception of the village itself, which faces south toward Lattimer over several miles of abandoned strippings. (1967a: 3-2)

Of the sixteen towns listed in the report, Pardeesville is reported as having the largest percentage of blighted dwelling units at 23 percent among an average of 6.6 percent for the region. Obstacles to redevelopment include stagnant and low home values, sewage and storm water disposal, water supply, rubbish storage, weed growth, and attractive nuisance and fire hazards (Kendree and Shepherd 1967a: 3-18–3-21). "Stagnation," closely tied to blighted conditions, is assessed in the "value of homes for modern use," either in the anachronistic construction or condition of buildings, the extended occupation of residents, and the number of facilities such as bathrooms and bedrooms (Kendree and Shepherd 1967a: 3-15–3-17). The problem of sewage disposal and water supply targets drainage that runs into ground surfaces and local streams including the wildcat sewers installed by residents in many villages. The report warns that outdated local systems of water delivery such as the water tank maintained by the Pardeesville Water Company create "a potential public health problem" in their lack of maintenance, water quality, and lack of chlorination (1967a: 3-19). The problems of rubbish and rubbish collection and disposal target the tendency of inhabitants to either dump refuse in open spaces or to collect items such as aging equipment and automobiles on property. The report notes that:

> Open burning is generally conducted on the lot. Often cardboard cartons, 55 gallon drums or even no containers at all are used for storage. Such sources are excellent breeding grounds for insects and rodents. In addition, discarded items are often maintained in old sheds and garages, contributing to the unsightly appearance of many rear yards. These are also an excellent attraction for rodents. (1967a: 3-19)

The report notes that at the time of writing there was no organized rubbish collection in the area. As a result, residents dumped refuse in open areas, strippings, or at local dumps.

The problem of weed growth, an issue around abandoned buildings and junkyards, "presents a pollen problem for persons afflicted with certain allergies" as well as serving as a breeding ground for insects through the collection of rainwater (Kendree and Shepherd 1967a: 3-20). "Attractive Nuisance and Fire Hazards" addresses the many ruined or abandoned dwellings, industrial structures, culm banks, strippings, and unregulated junkyards throughout the region. These landscape features are described as unsightly or as dangers to the unwary, namely, children, in the way of enticements risking collapse or fire. Applying a disease metaphor often associated with blight, the report suggests that these material factors multiply upon each other: "Any deteriorated conditions, in fact, encourage others to develop. In this way the problem gradually becomes more severe when no steps are taken" (1967a: 3-21). Among the solutions proposed by the comprehensive development plan is the strict enforcement of zoning laws as well as building, sanitation, and health controls (Kendree and Shepherd 1967b: 1-3–1-4). The enforcement of these controls is recognized as potentially valuable forms of township revenue accrued through various taxes, subventions, licenses, and other permits (1967a: 4-10–4-12, 5-5).

The authors of the report suggest, tautologically, that because the recent influx of suburban housing has already created additional infrastructural demands on the township, the future role of governance "will be to provide even more suburban housing." This development is justified by the need to produce the taxable income from home ownership necessary to fund the apparatuses of administration, maintenance, and enforcement in the township (Kendree and Shepherd 1967a: 3-14). Changing population density in and around patch towns, it is noted, will require constructing high-value, low-density housing in undeveloped areas at their periphery and altering preexisting high-density areas within the towns into a "secondary core" (1967b: 2-3). To accomplish the latter task, the report proposes that selective older dwellings be demolished to make way for open space, or traditional double houses be converted to "single-family detached dwellings" (1967b: 2-3, 2-10).

The "problem of local identity" for township government is cited as a barrier to redevelopment given the regional tendency to look upon the neighborhood as the basis of community. The report delivers a number of solutions, all of which reflect some form of social production: a machinic development of a new subjectivity supportive of, if not dependent upon, these new forms of infrastructure and governmentality. The authors argue, tautologically, that without a "closer relationship between the government structure and residents . . . it will be difficult to justify the underlying reasons for the comprehensive plan" (Kendree and Shepherd 1967a: 5-2). They clearly recognize the productive relationship between the intimate landscapes of the patch towns and the solidarity of local communities in their suggestion that addressing the housing problems of the township may "threaten the existence of the neighborhoods" (1967a: 3-18). Echoing the early–twentieth century work of Edward Ross, they report that new technologies may already have begun to solve the problem of insularity:

> Today the automobile and television have done much to change living habits. Commutation is relatively easy, and since coal mining is no longer as important to the local economy, there is little reason to continue to reside in the villages. In recent years, many persons who are moving into the new subdivisions do not have such ties. They tend to look more directly to the Township. (1967b: 1-8)

Last, the report encourages the development of neoliberal entrepreneurialism on the part of residents over their ties to historical community landscapes. The authors suggest that encouraging residents to assist with "planning improvements, elimination of eyesores, and assistance in the purchase of supplies for improvements" at the village level may encourage their support for governmental controls with the inducement of "increased market values" (1967b: 1-9).

A section on enforcement proposes several forms of governmental control over the built and social environment. This section endorses the tactical use of ordinances to discipline and control social behavior. Enforcing standards over the built environment entails three types: predevelopment, use, and construction controls. Condition and activity controls, however, regard policing the social environment (1967c: 2-1–2-1). Condition controls extend to the built environment through its social use and are controlled through health, sanitation, and housing ordinances. Condition controls are provided in the form of negative freedoms which apply to the legal co-responsibility of a general social realm. Examples include "freedom from hazards which may cause accidents; freedom from improper storage of rubbish and garbage, freedom from noxious weeds and objectionable plant growth," and "freedom from the depreciation of surrounding property" (1967c: 2-12). Although difficult to enforce given their slipperiness in the realm of law, the report deviously cites condition controls as "probably the most important tool in enforcement." They further suggest, "Greater imagination in writing legislation and effective enforcement is the key to its effectiveness" (1967c: 2-2). In essence, social behaviors resistant to traditional methods of coercive control can be policed if they can be spatially encoded in zoning.

By the 1960s, the once-bustling commercial center of Hazleton that was focused around Broad and Wyoming Streets was largely absent of commercial establishments. In time, the emptied husks of commercial buildings downtown were demolished by municipal condemnation justified by urban redevelopment (Tarone 2004: 127–169). The Hazle Township Redevelopment Plan proposed profiting, if not hastening, this exodus and deterioration of commercial activity by concentrating retail businesses in shopping centers on the periphery of the city. "Zoning, building and health controls" laws would be developed to encourage "coordinated . . . groupings of commercial activities" accessible mainly to automobile traffic (Kendree and Shepherd 1967b: 1-4, 2-4–2-5; 1967c: 1-2). Four major areas were proposed as major shopping centers or malls. The largest of these areas, was already "expected . . . to become in the future a large regional center representing a relocation of Hazleton core to the east" (1967b: 2-4). The Laurel Mall, developed in 1972, realized these plans and continues today to offer the city its largest concentration of national-brand retail. The demise of urban commercial retail in Hazleton and the deterritorializing of its role as a hub of community social life followed trends occurring in American cities across the nation throughout the first two decades of the post-war era (Cohen 1996, 2003; Hanchett 1996; Jackson 1996). As Lizbeth Cohen reports (2003: 257), suburban retail would outstrip that of urban trade in this period, making up almost 60 percent by 1961, up from 4 percent in 1939. The dependence upon these new retail centers cemented the parameters of a new spatial, material, and economic world based around mass consumerism and petroleum-based energy sources (Rose 1981; Mrozowski 2014). As Rose (1981: 96) reminds us, for "a city and a way of life built during the heyday of coal and rail," Hazleton's only choice at this time was to "adapt to the new energy source or suffer . . . even more massive depopulation and social disintegration."

BEFORE THE LAW

Pardeesville residents reported that the periods after the war and the passing of ordinances were transitional, and frequently tumultuous, years for the community. The changing material landscape is both cause and effect of these changes, many of them positive, for the community:

> DM: Things did change slowly before the war, but then after the war then they started moving ahead. The soldiers came back and they married and they started building houses and going to school. G.I. school.
>
> MR: So did the community change a lot when people moved different places? From being close together to being spread out some?
>
> DM: Yeah, people were starting to spread out the way it was. I mean because a lot moved out of town. There was no work here. A lot of them moved to New Jersey, to work in factories in New

> Jersey. And then they started CAN DO. That helped the area out. But a lot of them were out of town. And then. . . . people started becoming more independent as years went by, they became more and more independent. And they were separated, spreading out. . . . like today, what you see today. Sometimes you don't even know your neighbor, this town here, there are a lot of people I don't know and at one time I knew them all.

Residents and newspapers reported that beginning around this time neighbors fought among themselves over the present and future of the community.

By the 1970s, even the water company, the first example of communal self-organization, was falling apart. It was the victim of aging infrastructure, mismanagement, and a community reluctant to pay into traditional informal modes of organization. A resident commented:

> The company was in a shambles. The records weren't up to date. They were frivolous . . . overinflated. . . . people [did]n't want to pay for their water. It was like pulling teeth. I was going house to house collecting delinquent dues. Where you going to get all this money to do this and that? The lines were getting all deteriorated. They fought. (MD, interview, March 21, 2013)

Further atomization of the once-intimate communities of the coal patch towns is exemplified by the lawsuits between residents beginning in this period. Neighborhood disagreements were often framed in the language of health, safety, and sanitation ordinances, but centered upon the protection of individual property values. In April of 1967, a resident of Hazle Township by the name of John Sacco petitioned township officials to prevent neighbors from burning garbage next to his property and force them to remove an unsightly shanty (*Standard Speaker* [*SS*], April 28, 1967: 2). Sacco is quoted in the newspaper petitioning as follows: "I live in a residential area," he said. "Why should I and my neighbors with well-maintained houses have to put up with an unsightly structure?" he asked. Sacco returned to the township's public meeting in May to request that the municipality use the nuisance ordinances to remove the shanty that was, in his opinion, both a "health hazard" and an "eyesore" (*SS*, May 31, 1967: 15). Sacco argued his case by bringing up the examples of recent buildings in Hazleton city that had been demolished by the Hazleton Redevelopment Authority and that had also been declared as "eyesores" to the community (*SS*, May 31, 1967: 15).

On the 28th of April, 1967, soon after the publication of the *Hazle Township Comprehensive Plan*, the local newspaper announced a township drive to eliminate nuisances throughout the municipality. Planned areas of enforcement included the illegal dumping and burning of garbage and the parking of junked cars. A captioned photo in the *Standard Speaker* from May of 1967 depicts a junkyard in Pardeesville, noting that cars, "have been tagged or their owners have been warned . . . if these cars are not moved by Monday, the township will have them moved away at the owners [*sic*] expense" (*SS*, May 5, 1967: 6).

In July of 1967, the township passed extensive nuisance and building ordinances and looked toward creating a police force to ensure compliance with its new laws (*SS*, July 27, 1967). Ordinances passed at this time included a law against the accumulation or dumping of garbage, and the growth of weeds to a height of more than six inches; a law prohibiting "the carrying on of any offensive manufacture or business"; and a justification for the township to remove "any nuisance or dangerous structure on public or private grounds after notice to the owner to do so." The newspaper quoted the law as stating that "offensive manufacture or business" includes "any enterprise of business which constitutes a nuisance 'or is detrimental to the safety, health and welfare of the community'" (*SS*, July 27, 1967: 13). A promise was made to "vigorously enforce" the laws.

In December of 1967, the Hazle Township Zoning and Planning Commission received a series of petitions at a public meeting from residents charging that enforcement of the ordinances was "discriminatory and unfair" (*SS*, December 5, 1967: 15). In one instance residents complained that a trailer placed somewhere in the vicinity of Church Street in Pardeesville was given a building permit despite exceeding by two feet the length permitted. The resident, who resided in the dwelling investigated in the present study, could see the trailer from her residence and protested that no building permit was publically exhibited at the site as required by the ordinance. A second complaint revolved around a cow one family kept despite the area's residential zoning designation banning livestock. A member of the twelve-person group suggested that "the zoning officer wants everything to stay as it was 20 years ago."

They continued, "If this is so, then let us pay taxes as they were 20 years ago. Give the money back to the people that paid for zoning and building permits, as the present system is a farce" (*SS*, December 4, 1967: 14; December 5, 1967: 15). On the 13th of April (*SS* 1970: 13), a resident of Pardeesville charged that the township was failing to properly enforce antiweed ordinances in the town, claiming that while one side of a double house was occupied with a well-tended garden, the other half remained vacant and full of high grass and weeds.

The increasingly bureaucratic demands residents made of each other throughout the last few decades of the twentieth century suggest that, at least tactically, they adopted the language of renewal and management as understood within the framework of individual rights or negative freedom. John Sacco and others already recognized that condition controls structured as negative freedoms unambiguously positioned the legal rights of individuals over traditional or unwritten community values. This was particularly the case when individual rights to things such as privacy, profit, or privilege were considered to be integral to broader political or economic ideologies.

In this process, aspects of the historic landscape of the patch town—gardens, population density, architectural aesthetics, privacy or its lack thereof, animal tending, refuse disposal—each took on an altered significance within the context of economic renewal. Ethnographic accounts suggested that heirloom plants brought over from Italy, including garden vegetables and fruit trees, overgrown or untended, were removed as weeds along with the knowledge of how to cultivate them. The same was true of domesticated animals. Collections of machinery consisting of automobiles and other scavenged material—the source of much banked capital, surplus parts, technical repositories, and pleasure—were removed as junk. Finally, the intimate spatial configurations of the town that both nurtured and demanded informal and communal methods of arbitration and communication were forever altered. A new seamless regime of structural and symbolic violence had taken hold as a softer form of social control.

THE COMMODIFICATION OF COAL HERITAGE

The authors of the *Hazle Township Comprehensive Plan* conceded that many of the problems of the area were linked to the coal industry, and therefore, "no real progress can be made without acceptance of this past in a manner which can be a part of the future development of the municipality" (Kendree and Shepherd 1967b: 1-4). The plan authors recommended that one of the solutions to the region's economic problems would be to introduce tourism to the region. To provide a spectacle sufficient to attract tourists from the wealthier cities of the East, they recommend that little to nothing be done to mitigate the ruined landscapes of the coal industry:

> The emphasis must be on the starkness of the area. A site to be developed for this purpose should not be planted and emphasis must be placed on maintaining this visual impression of lifelessness. Sometimes the emphasis can stress vast expanses of dissimilar surfaces, such as the jumbled rock of a high wall of stripping counterbalanced with a smooth water surface of an inaccessible lake . . . tourism is the only practical solution to the failing local economy as it alone would not require the massive funds necessary to clean up the wasted and ruined landscape. Instead it would turn them, along with an anachronistic social life and infrastructure, into a productive spectacle. (1967b: 5-3)

The authors drew from an essay by economic geographers George Deasy and Phyliss Griess (1961) who asserted that the spectacle of industrial ruin in the region "is bizarre and grotesque, and to some visitors even repulsive, but it is nevertheless mesmeric in quality." The landscape represented "awe-inspiring . . . earth manipulation" of gigantic proportions for which "engineering feats such as the Suez and Panama Canals pale in comparison" (1961: 3).

Deasy and Griess recommended that the people of the region, in combination with the anachronistic built environment, should be part of the preserved spectacle. While the region is hardly "the most pleasant site in which to establish permanent residence . . . the aura of the past that permeates these towns can have a definitive appeal" to the tourist from modern suburbs who would recognize "the tempo and atmosphere . . . reminiscent of a past generation" (1961: 4). Such things as "the occasional appearance of a horse-drawn peddler cart" or "neighborhood men in aprons gossiping in the yards" would offer the modern urban visitor "a page from his own life," or "earlier years in a world that moved at a slower pace" (1961: 4–5). Moreover, the retention of ethnic migrant cultures in the form of food, dress, and religious architecture, they suggested, offered much to stimulate tourist trade. To "capitalize on the region's distinctive . . . atmosphere" (1961: 7–8), they proposed that "efforts should be made to preserve" some of "the region's urban centers in their present state" rather than "attempting to modernize them" (1961: 5). Such deliberative preservation should be accompanied by the development of specialty food, dress, and handicraft shops, "appropriate redecoration and landscaping," and "the deliberate use of more horse-drawn vehicles" (1961: 5).

Deasy and Griess (1961: 7–8) warned that there was much work to be done in the way of producing the infrastructure of travel, hospitality, and services necessary to support the needs of out-of-town guests. Potential tourist sites would require signage, facilities, and showmanship. National advertising campaigns would be needed to regiment the public's opinion of the region in the direction of curiosity and interest. Moreover, they contended, tourism would require social production in the form of "major innovations in the mental orientation and attitudes of regional and community leaders, and significant changes in some of the physical attributes of the region." The advantage of introducing tourism to the region as opposed to any alternative industry was that it alone would not require the herculean effort or investment capital needed to alter the ruined environment (Deasy and Griess 1961: 3). In fact, the catastrophe of the coal industry and the anachronistic behavior of its isolated residents, within appropriate constraints, would be the very foundation necessary to "reestablish the basis for renewed accumulation" (Harvey 1985: 116).

The *Hazle Township Comprehensive Plan* (Kendree and Shepherd 1967a, 1967b, 1967c) and "Tourism for the Anthracite Region" (Deasy and Griess 1961) approached the economic and labor conditions of the region from different angles, but with a similar goal: to recapitalize upon what remained of a region in major decline. These planners mandated an approach integrating the vertically integrated governmentalization implied in the urban renewal movement. As for the aging, overspecialized workforce displaced from the mining industry, they ambivalently suggested that it was no longer "enough to simply provide work for the unemployed, but to start a development program" (Kendree and Shepherd 1967a: 5-2). Although they did not state it, perhaps they imagined that unemployed miners could perform the touristic spectacle of "old men seated on benches in front of neighborhood stores" as recommended by Deasy and Griess (1961: 5)? For Deasy and Griess, a short-sighted utopian plan appropriated the profitability of the ruins of social and material life in the anthracite region. In a sense, their plan was far less ambitious in that they proposed only the replacement of one monolithic industry for another, tourism for coal, which would occupy the same sites and retain the same retarding and isolating factors.

THE CAN DO PHILOSOPHY AND THE RISE OF FLEXIBLE ACCUMULATION

The Hazle Township Comprehensive Plan is structured in the national discourse of renewal. Its endeavor was to recolonize the landscape for new forms of postwar economy. Despite the impression of rigorous data collection, the study neglected a matter of key importance to the region: the massive unemployment resulting from the end of the mining industry. Instead, it focuses on creating a selected workforce of white-collar labor to stimulate tax revenue. While forward-thinking, the study overlooked the remnants of the past, as if it were possible to accelerate the present to forget the people, history, and material landscapes.

In truth, the region already had an early, relatively successful redevelopment effort known as CAN DO, which mobilized as many residents as possible to finance the redevelopment of reclaimed mining sites for new industry. To accomplish this, CAN DO cleverly capitalized on the sometimes-paradoxical spirit of communal support and competition that defined the dynamics of the region. A self-historicizing work produced in 1974 reflected on this strategy:

> The most important ingredient in community action is the ordinary people of the community. The people of Hazleton came from many different countries to the rugged mountains to work, to earn a living, to make a new life. They have a strong sense of individual family responsibility. They lived by a creed of "doing for" their family and community as constructive citizens and looked upon anyone who lived otherwise as somewhat of a slacker and parasite. (CAN DO 1974: 41–42)

In the early 1950s, repeated failures to court industrial prospects to the region revealed the inadequacies of the ruinous mining landscapes.

As early as 1955, the first CAN DO drive, Operation Trees, targeted these aesthetic issues, operationalizing community volunteers and corporate and state-sponsored donations. CAN DO organized the planting of trees donated by corporate sponsors along roads throughout the region to shield the grotesque piles of mining waste from the view of passers-by (*Plain Speaker*, May 26, 1955: 13). As late as 1971, the newspaper reported an unlikely coalition of boy scouts, representatives of organized labor, and representatives from the Pennsylvania Bureau of Mines and Pennsylvania Power and Light planted 4,000 trees along the CAN DO freeway to "improve the environment, produce oxygen and give the Scouts the opportunity to work on a worthy program (*SS*, April 27, 1971: 20).

The first DIME-A-WEEK drive conducted in 1955 asked all residents of the town to donate 10 cents a week to fund the development of the first building shell for new industry (CAN DO 1974; Rose 1981: 113–115). Lunch pails were placed throughout the city to make the campaign as readily visible as possible. In a public relations move worthy of Edward Bernays, the local media stimulated this effort by daily publication of the names of contributors and businesses in the newspaper to inspire competition. The CAN DO history explains that "the listing of names. . . . exerted subtle pressure on those who had failed to donate. No one wished to be placed in the position that he failed to help his family, neighbors, or community" (CAN DO 1974: 42). As the CAN DO organization depended entirely upon volunteer base, the public campaigns gave residents an active role in aiding the community. Thus, the core community values of self-sufficiency would not be put at risk in the welfare efforts of redevelopment. The fundraising efforts were, by all accounts, an incredible success that often raised funds well beyond projected goals. The result was a revolving fund for repeated action (Dublin and Licht 2005: 127).

By the mid-1950s, CAN DO had purchased reclaimed mining land to the southwest of the city to produce an industrial park. They installed sewer, water, and electrical infrastructure to these sites. Between 1956 and 1974, the corporation added a total of sixty new or expanded industrial firms to the employment base of the city. These companies provided over 7,000 jobs, amounting to earnings of nearly $50 million for the local economy (Dublin and Licht 2005: 117). These jobs were largely in unskilled trades such as light manufacturing, including garment and textile industries, metal trades, and assorted plastic- and polystyrene-based fabrications (CAN DO 1974: 75–77).

Retrospectively, the efforts of CAN DO were dependent upon, or more precisely, nested within, a national economy in transition (Dublin and Licht 2005; Rose 1989). David Harvey posits the energy crisis of 1973 as the end of Fordist economics and the dawning of a new age of flexible accumulation. Preceding this is a decade or so in which the "inability of Fordism and Keynesianism to contain the inherent contradictions of capitalism became more and more apparent" (Harvey 1990: 141–142). This new financial logic coincided with the latter half of the redevelopment efforts in the region. Two major characteristics of this new economy deeply affected redevelopment efforts in the region: (1) the intervention of the state in regulating the economic landscape upon which capital accumulation occurred, and (2) the expansion of American industry's capacities—technological, logistical, and economic—for the rapid liquidation and movement of capital and labor markets. In effect, these developments split in half this period in the region's history between the postwar period and the dawning of regimes of flexible accumulation toward the end of the 1960s.

Dublin and Licht (2005: 126, 239) suggested that the distribution of federal and state assistance capital in the region followed two patterns that corresponded to this split. The early period reflected regional efforts to redistribute funds based upon genuine need. As growing professionalism in fundraising at the regional level intersected with an important legislative decision regarding laws governing the eligibility of recipients, capital began to flow into increasingly concentrated pockets of industry and place. A review of fund distribution reveals that this change occurred roughly around 1968 (Dublin and Licht 2005: 126, 239). According to at least one assessment, by 1979 the federal and state assistance programs that played a major role in redevelopment in the region served to make the "rich [get] richer and the poor [get] poorer" (Dublin and Licht 2005: 126).

Flexible accumulation is characterized by regional competition for work occasioned by the presence of newly exploited, less-expensive global labor markets for industrial work. Industry accomplished liquidity of capital with the facilitation of the "permissive technologies" of mass transportation and mechanization that allowed entire industries to relocate practically overnight (Bluestone and Harrison 1982: 115;

Cowie 2001). The success of the community efforts to bring industry to the region hinged upon this dynamic. The incredible competition between communities across the country to court industry led to the undercutting of stability and quality in the work that could be captured and integrated. CAN DO and other regional community development firms succeeded by a combination of enticements, not the least of which was the large base of unskilled, unorganized mining labor. These workers were, for the most part, in desperate need of work. The ostensible draw of this labor pool was that they would require only minimal wages with a decreased risk of absenteeism (Dublin and Licht 2005: 132–135). On top of this, the monetary enticements offered by a variety of federal and state assistance grants and community donations were sweetened by introductory periods of tax-free occupation. In many cases, companies left when their time was up (Dublin and Licht 2005: 132–135). Therefore, any lasting results of redevelopment efforts were mixed. At best, they offered the region a second wind based upon conditions that seem nearly epochal in the region's economic history. The surplus labor army of men and women of the region were offered a place in a competitive global market for precarious, low-wage work. This time it was not only for the excepted migrant workers, but for all the region's working population.

CONCLUSION

In Chapters 5 and 6, the actions and relationships between three main social groupings in the anthracite region are described: community members, local government, and the business community. Conspicuously missing are the feudal elites of old—the coal operators who left the region to diversify their portfolios on other interests, abandoning their failing investments. In the period after the Second World War, the region's working, merchant, and professional classes found themselves alone to salvage the life of the region as best as they could. In truth, defining a strict division among these groups by the end of the 1960s is difficult. By this time, residents of Pardeesville and Lattimer served in local government and within the business community, pursuing the recolonizing of their own small worlds. In fact, a construction company based out of Pardeesville was awarded some of the largest contracts to demolish buildings in South Hazleton during the period of renewal (*SS*, February 11, 1969: 15).

A fragment from the Walter Benjamin text that begins this chapter illuminates the self-annihilation that characterizes "the Destructive Character":

> The destructive character is young and cheerful. For destroying rejuvenates, because it clears away the traces of our own age; it cheers, because everything cleared away means to the destroyer a complete reduction, indeed a rooting out, out of his own condition. (Benjamin 1978: 301–303)

This quote exemplifies perfectly the spirit by which those responsible for renewal efforts justified the destruction of historic landscapes. These landscapes had come to represent, in consonance with the language of renewal, the breeding grounds of corrupt, unhealthful past practices and social behaviors. In a way, the destructive period of the 1960s was an effort by those who remained in the region to abolish the memory of their own past, pursuing the most immediate and lucrative means to "the way leading through it."

The redevelopment efforts of the 1960s amounted to an effort not only to "recolonize" (Rose 1989), but rather to reclaim the landscape and social relationships of the anthracite region. The effort to restructure the ecology of the region into a productive landscape for capitalism took three forms. First, they sought to recapitalize the region's resources that were in great supply after the exit of the coal companies: labor and real estate. The latter required a material transformation into aesthetic and infrastructural landscapes attractive to new industry. Ordinances transformed the settings of the everyday (architecture, weeds, refuse, water, air, space, and privacy) into commodities subject to lawful appraisal and enforcement. A corollary of this, outlined in *The Hazle Township Comprehensive Plan*, is an effort to cultivate residential areas as tax-producing resources, which would in turn maintain the governmentalizing apparatus needed to regulate and administer to colonizing efforts.

The withdrawal of the coal industry left the region burdened with a large force of aging, unemployed, unskilled, and unorganized workers (Rose 1981). Despite the lack of employment, strong familial and community ties bound many individuals to the unique landscapes of the region. By the 1970s, this labor market perfectly met the growing demands of a global economy moving toward flexible labor regimes and away from Fordist models of stability, institutional support, and paternalism (Harvey 1989: 150). The once-extensive organizational power formerly held by the UMWA over regional industry was impotent to protect workers divided into ephemeral microindustries. The same was true of industries utilizing nontraditional labor arrangements such as part-time, temporary, or subcontracting to maintain the liquidity of operations (Harvey 1989: 150).

As before, this instability was mediated by the surplus enjoyments of life in the region: community and family support, gardening, opportunistic scavenging, and informal work arrangements. However, these benefits were subject to diminishing support by a community comprised of increasingly atomized individuals adopting entrepreneurial attitudes to space, labor, and governance. Material aspects of the space, which amounted to so much more than mere aesthetics in previous eras, were recoded as nuisances and removed. At stake was the capacity for the materiality of landscape to remember, reproduce, and channel social relations in a manner responsive to the exigencies of uncertain economy.

8

Conclusion

> Truth is not—as Marxism would have it—a merely contingent function of knowing, but is bound to a nucleus of time lying hidden within the knower and the known alike. This is so true that the eternal, in any case, is far more the ruffle on a dress than some idea.
>
> *Walter Benjamin (1999: 463)*

Great social and material progress, human achievement, and communicative growth mark the trajectory of the twentieth century. Ironically, it is also a time of great inequality, violence, social fragmentation, and nostalgia (Hobsbawm 1994). In our present, the contradictions of our recent past have yielded social tensions that seem indissoluble if not simply inexplicable. Without an accounting for their historicity, it is difficult to imagine their resolution or posit their trajectory. These contradictions haunt our present through a dark heritage suffused in cognitive dissonance. This study is an endeavor to find the material grounds of an archaeology of the twentieth century that accounts for its contradictions.

Early in the twentieth century, Walter Benjamin reflected upon the complexities of encountering truth in history. He asserted that finding truth is not simply a function of comprehensive historical knowledge—it demands a confrontation with our very methods of learning about and knowing the past. He does not leave us stranded without a method or object of study. Benjamin posits a historicity that foregrounds material (that is, archaeological) modes of knowing. As he asserts, material change (that is, "the ruffle on a dress") is the fundamental ground upon which historical truth, the "prehistory of the historical moment," may be discerned (Buck-Morss 1989: 287–330).

Observing the rise of fascism and totalitarianism in Europe, Benjamin sought its origins in the aesthetics of past eras. In the suppressed memory of the recent past is the unconscious source of desire in the present. In the context of mass production, this forgetting is materialized in the fashion cycle, through the constant fleeing from "the outmoded" (Benjamin 1978: 181). In this way, objects and ruins materialize the suppressed antagonism or gap between the aims of technological or social progress and the lived material realities of the present. In the materiality of the mid–nineteenth century, he could already see that class struggle and inequality operated both in contradiction, and in contingency of the excessive material surpluses produced by the steady acceleration of mass production. The chief products of these factories, Benjamin found, were alienation and desire. Ironically, it was this very same material overproduction, with its promise of material comfort and accumulation for all, which also holds the idealized solution to the problem of inequality.

Benjamin's historical and material project inspires the broad structure of this study. I use archaeological methods to pursue three interlocking inquiries fundamental to understanding the contradictions and ironies of the twentieth century. The study generates context and analysis both regionally specific and generally applicable to a nationally significant archaeological context. At its broadest application, this work is an endeavor to produce the outlines of a nuanced periodicity of the structural violence of everyday life in the twentieth century.

The first inquiry investigates the role of structural violence in objective settings. These objects of analysis include the systemic and symbolic violence of landscapes and semiotic systems of racialization in justifying or triggering moments of explicit subjective violence. Each are distinctly visible in the archaeological record as the sum of the visible features within the productive landscape of the coal company towns. This inquiry begins with an investigation into the contexts of the Lattimer Massacre of 1897 in Chapters 2 through 4, to assay the role of these landscapes in the region's violent history.

The remainder of the book is devoted to tracing the trajectory of the town in the interim between turn of the twentieth century and the present. The second inquiry arises out of an analysis juxtaposing the discrimination of immigrants in the past with that of the present. The archaeological account is once again a matter of historicizing trajectories. Archaeological materials, in combination with archival sources, trace the trajectory of historically excluded groups into contemporary neoliberal subjects. In other terms, this inquiry considers how historically oppressed groups can become complicit with oppression later in history.

The third inquiry historicizes the development of contemporary forms of structural violence, contrasting them with moments of explicit violence at the end of the last century such as that of the Lattimer Massacre. On a broader scale, I examine the development of these soft forms of social control and coercion across the *longue durée* of the twentieth century, focusing on the development of vertically integrated economic and governmental structures such as mass consumerism, neoliberalism, and governmentality, which serve to stabilize the social antagonisms of the past. I pursue these questions beginning in Chapter 5, commencing with an examination of social, material, and industrial changes in the shanty enclave across the twentieth century. This landscape is traced through the materiality of machinic mass consumerism in mid-century, a period of postindustry in the 1950s, urban renewal of the 1960s, and the rise of neoliberal subjectivities and economics in the 1970s.

THE STRUCTURAL VIOLENCE OF EVERYDAY LIFE AND ITS ROLE IN JUSTIFYING EXPLICIT VIOLENCE

Historically, the anthracite coal industry was interminably linked to broader historical contingencies. Global depressions, political developments, and mass migrations shaped the social and economic patterns in the region, accruing as exaggerated reactions to the circumstances of a fickle industry. Regular boom-and-bust periods ensured labor and capital were constantly at odds. The region often served as the first stop in the proletarianization of waves of peasantry freed from feudal conditions in Northern and, later, Southern and Eastern Europe. By the late nineteenth century, large numbers of immigrants from Eastern and Southern Europe sustained the ranks of an army of surplus labor, joining or slowly replacing earlier generations of nativized workers. This pool of surplus labor granted capital the logistical flexibility to respond to the exigencies of an unpredictable market.

Company, or patch, towns were the foundational landscape features of the Anthracite Coal Industry of Northeast Pennsylvania. Capitalist engineering of space serves not only to concentrate production to minimize costs, it also serves the productive function of differentiating and reproducing a striated labor force, physically (that is, systemically) and symbolically, in a manner conducive to productivity (Harvey 1989: 111). In time, the coal company towns of the anthracite region came to reflect the racialized labor-class hierarchy that prevailed throughout the late nineteenth and early twentieth century.

The shanty enclaves functioned as autonomous and unordered space within the margins of the company town, resisting its order and visibility, although essential to the broader system of its productive function: providing the colliery with a steady supply of surplus labor. Archaeology and archival research suggest that far from being ephemeral, these settlements were constructed fewer than fifteen years after the initial founding of the towns, and remained a permanent, albeit shifting, fixture on the landscape. These findings are quite different from scholarship that interprets such satellite communities as alternative spaces for those seeking to operate outside of capitalism (Goddard 2002; Hardesty 1998; Matthews 2010: 138–141). Moreover, recognizing the central role of precarious labor surpluses throughout the long history of capitalism can illuminate its role in the present. Indeed, the threat of precarious labor conditions is today no longer restricted to the proletariat, but is a fact of life for many in the present, including immaterial and intellectual laborers (Lazzarato 2012).

Middle-class visitors to the shanty enclaves reflected upon the manner in which its aesthetics expressed the racialized inferiority of its inhabitants. In examining the organic quality of shanty architecture, one observer was moved to note that the builder was "a laborer, a struggler for mere existence, not deft in the use of tools, nor with an eye for the symmetrical, nor with an appreciation for anything beyond the most primal facts of living" (Hambidge 1898: 825). To the middle-class businessmen that were deputized to defend the colliery of Lattimer in 1897, the differentiation of a racialized order by architecture, space, and consumer goods was apparent. Moreover, such observers perceived this built environment as attributable only to the immigrant laborers, in isolation from the broader historical context that contributed to its character. These authors were entirely ignorant of the structural role the mine owners had in engineering its existence and maintaining its conditions to their advantage through regimented racial hierarchy, poverty, and debt. Moreover, they could not see the dynamic character of the enclave, which had already been in a state of constant improvement and adaptation by its long-term residents for over twenty years. For these authors, this place had no history and no future.

Company towns may seem like a *monadic* form: unreducible, totalizing, dominating, and static. However, when viewed in the *longue durée* they reveal a shifting dynamic of violence, contradiction, and movement. Archaeology, coupled with digital cartography, reveals a seriation to its innumerable transformations. Archaeological investigations of a shanty house reveal an adaptive organic architecture. Moreover, archaeological findings challenge the notion that the shanty reflects an ahistorical static expression of biological character, or an ephemeral or nomadic occupancy by migrants. Rather, archaeology reveals a long history of active, committed efforts to create a healthy, safe, and intimate communal environment even within the harsh living conditions offered by the coal mining industry.

In the years leading up to the event known as the Lattimer Massacre, the mine operators and the middle class of Hazleton benefited from stable profits issuing out of industrial-scale coal extraction. These profits depended upon mechanization and labor exploitation. This privilege afforded them comfortable, stable domestic lives, isolated from the precarity, environmental degradation, and danger of the company towns. Threatened only by unsteady markets and the increasing organization of longer established craft labor, the industry turned to a combination of mechanized production and unskilled immigrant labor. For a time, this underclass of surplus labor was far less demanding of stable wages, and costly safety measures, than earlier generations of nativized labor. Systems of objective violence differentiated these populations through domestic landscapes and racialized character, justifying a split labor force. When these new immigrants sought rights equal to those of earlier generations of workers, the stability of these arrangements were called into question. The violent response by those who profited from this arrangement was swift and unequivocal.

As an instance of explicit subjective violence, the Lattimer Massacre transpired over a period of only about three minutes. Its import is in the way it illuminates how everyday life is punctuated by forms of everyday violence. During the Lattimer Massacre, the power over life and death hung delicately in balance in the hands of an armed posse of middle-class businessmen. From their perspective, charged with maintaining the safety and stability of society, this was sufficient excuse to elicit their lethal response.

FROM MIGRANT LABORER TO NEOLIBERAL SUBJECT: OR, HOW THE HISTORICALLY OPPRESSED CAN BECOME COMPLICIT WITH OPPRESSION

At the end of Chapter 4, I proposed that the shanty enclave at the edge of the company town was not merely an ephemeral settlement organized as a historical contingency. Rather, it was a fixture that was central to its functioning from the beginning of its existence. As I described in Chapter 4, its execution was an act of symbolic violence differentiating portions of the labor force as surplus. Viewed in the *longue durée*, the shanty enclave models for all workers the presiding conditions of precarity and self-sufficiency demanded of all in late–twentieth century capitalism.

Early–twentieth century politicians, scholars, and social workers pinned the enfranchisement of America's new immigrants from Eastern and Southern Europe to their adoption of mass consumer habits (Cohen 1986). Importantly, these observers were not apprehensive about the satisfaction of immigrants' creaturely needs. Rather, their concern was with the kind of surplus desire that is patently unappeasable, what one observer of the anthracite coal region (Roberts 1905: 220) defined as the "margin between [the] real wants and felt wants." An examination of consumer waste from an enclave of immigrant laborers living at the periphery of a coal mining company town in Northeast Pennsylvania reveals that indeed, by the middle of the twentieth century, even the inhabitants of this modest setting would be entangled in the materiality, practices, and networks of mass consumerism.

Chapter 5 considered the motivations and effects of machinic mass consumerism on the nation. In the period after the First World War, capitalists, intellectuals, government administrators, and advertisers proposed developing intensive forms of consumerism among the masses as a parsimonious but diabolical solution to two major crises confronting the nation: one of a financial nature, the other a matter of unstable governance (Roller 2015). The first crisis entailed the need to meet unmet demand for the surplus of commodities flowing out of mechanized industrial production. The second crisis, one of sovereign power, concerned growing anxieties over the nation's capacity to maintain control over its new immigrant population and the radical elements within the labor force. Literature and discourse from the time evoked the hope or belief that the distribution of material abundance, or merely the unquenchable desire for its consumption, would ameliorate, conceal, or divert attention from the material conditions that lead to struggle by radical and unassimilated elements of the working class.

An artifact assemblage from the mid-twentieth century recovered from Church Street in Pardeesville reflected changes in personal and household choices in consumption in the manner of individual commodities and aesthetic choices. By the second quarter of the twentieth century, the landscape of the enclave had radically transformed, coming to resemble a typical American suburb in architecture and spatial distribution. Thus, its origin as an ephemeral settlement of the marginalized was gradually erased. Consumer goods and other items found in the enclave at the time reflected the character of the mass consumer democracy instituted throughout the country.

As discussed in Chapters 5 through 7, by mid-century, spatial and architectural characteristics that had once clearly differentiated the shanty enclave from other company homes were erased or downplayed (Roller 2015: 365–407). The collapse of the industry brought big changes to the region in the years after the Second World War. Briefly, the town of Pardeesville flourished during an emancipation from the company when it divested itself of the town. As it did during its formative industrial period, this instability was mediated by the surplus enjoyments of life in the region: community and family support, gardening, opportunistic scavenging, and informal work arrangements. However, these were subject to diminishing support.

These changes corresponded with a national context of wealth and stability afforded by the postwar economy. Kazin (1995) argued that in the period after the Second World War, populist politics took a rightward shift, largely driven by white middle- and working-class

Americans who, through postwar prosperity, material abundance, federally aided home ownership, and mass advertising, had come to think of themselves primarily as citizen consumers. This right-leaning populist politics reflected the interests of a group far less willing to risk financial stability for radical change and were, therefore, far more accepting of the interventions of corporations in their lives. Furthermore, these groups grew increasingly suspicious of the expansion of government and its incursions into certain aspects of everyday life such as the burden of taxes (Oshinsky 1995). Kazin argued that by mid-century this remarkable shift had already produced a white populist politics that eschewed countercultural protest in exchange for an agenda designed to "thwart and reverse social and cultural change rather than promote it" (1995: 4).

As discussed in Chapter 7, urban renewal and other late twentieth-century developments, including a brief period of postwar growth and stability, removed the communal arrangements and adaptations that benefited the community during its industrial period. Juridical structures, governmental identities, and negative freedoms gradually replaced these local institutions and practices. Ultimately, the traumas of the recent past produced an alienated community of autonomous individuals networked to regional governance. The local economy was devastated a second time by neoliberal economic policies in the late 1970s. At the time, the global economy shifted toward the adoption of flexible labor regimes and away from Fordist models of stability, institutional support, and paternalism (Harvey 1989: 150). The once-extensive organizational power formerly held by unions such as the UMWA over regional industry was impotent to protect workers divided into ephemeral microindustries. The same was true as industries turned to nontraditional labor arrangements, such as part-time, temporary work or subcontractual labor, to maintain the liquidity and flexibility of post-Fordist operations (Harvey 1989: 150). Archaeology reveals that changes in the landscape and material culture of the region during this period of redevelopment removed the community's capacity to recall, reflect, and channel social relations in ways that could respond to the exigencies of increasingly uncertain economy. As discussed in Chapter 7, the former company town landscape was transformed into a tax-producing suburb cultivated to financially support local governance. The logic of governance replaced the communally negotiated political structures that the community had depended upon for nearly a century.

In the early 2000s, Central American and Caribbean immigrants began migrating to Hazleton, attracted by the work and low cost of living. Tensions began to rise between the local residents and the newest immigrants, associated with the growing national concerns about border security that arose in the wake of 9/11. In 2006 Hazleton was the first municipality to pursue laws aimed at limiting immigration. Although the laws, known as the IIRA, were never enforced and were ruled unconstitutional by the Supreme Court in 2013, the normalization of anti-immigrant discrimination immediately altered the character of interactions between long-term residents and recent immigrants. Through discursive practices of politic demagoguery and media, recent immigrants were racialized as illegal, unassimilable, lazy, criminal, and stubbornly refusing to learn English (Dick and Wirtz 2011; Kobialka 2012; Longazel 2013a, 2016; Longazel and Fleury-Steiner 2011). Ironically, the same globalized neoliberal ideology of efficiency that made the region and its workers unable to compete with undocumented migrant laborers for work also produced the economic conditions that justified their restriction by locals (Calavita 1994; Longazel and Fleury-Steiner 2011). Residents who support immigration restriction today target immigrant groups for their racialized inability to adapt to the presiding neoliberal values of entrepreneurship, self-sufficiency, compliance, individual striving, and freedom from collective enterprise.

The shantytown can be grouped within the categories of places Agamben (1998) describes as "the camp." As panicked (and yet engineered) responses to the challenge of diversity brought on by mass immigration, these places represent primal materializations of the political culture of the twentieth century. As the excluded who define the boundaries of the nation-state, the camp also signifies the consolidation of the national body through the intensification of governance. It is Agamben's (1998) suggestion that today there is a dispersal of this order. This dispersal results not from an elimination of contradiction by the cancellation of the boundaries between the excluded and included, but by a state of exemption that has become extended to all. Indeed, labor precarity today is no longer restricted to the proletariat or the laboring migrant, but is a fact of life for all workers, including immaterial and intellectual laborers (Lazzarato 2012).

THE DEVELOPMENT OF SOFT FORMS OF SOCIAL CONTROL: THE VERTICAL INTEGRATION OF NEOLIBERAL ECONOMICS AND GOVERNMENTAL SUBJECTIVITIES

The history of the twentieth century is the history of the subject in transformation. If we look past the content of traditional narratives of twentieth-century history, of ideological clashes, binary oppositions, influential actors, and events, a different material trajectory takes hold. If we focus analytical inquiry upon the nuanced effects of the politicization, financialization, and administration of everyday life, a different history emerges. Agamben (1998: 121) advises that such an exercise begins with observing the "contiguity between mass democracy and totalitarian states":

> It is almost as if, starting from a certain point, every decisive political event were double sided: the spaces, the liberties, and the rights won by individuals in their conflicts with central powers always simultaneously prepared a tacit but increasing inscription of individuals' lives within the state order, thus offering a new and more dreadful foundation for the very sovereign power from which they wanted liberation. (1998: 121)

Because this history concerns the ontology and relationships of the finest scales of the material lives of individual bodies as they articulate with the broadest scales of social organization, this is a subject for archaeologists.

The complex relationship between subjectivity, consciousness, and material objects has been convincingly explored in archaeological scholarship (Leone 2005; Leone and Shackel 1987; Matthews 2002, 2010; McGuire 2002, 2008; Sayers 2003, 2014; Shackel 1993; Wood 2002; Wurst and McGuire 1999; Wurst 1999). These archaeologists have convincingly illuminated the fundamental relationships that exist between individual and collective subjectivities and materiality. They also detail the vulnerability of this dynamic to the manipulations of power. Chapters 5 through 7 expanded upon this body of research in examining the development of new forms of subjectivity and consciousness during the second half of the twentieth century. Specifically, I addressed the development of vertically integrated forms of neoliberal economy and governmental subjectivities through archaeological examples.

I introduced the concepts of neoliberalism in Chapter 1 as a change in the grounds and effects of normative ideology in which financial logic has penetrated nearly every aspect of everyday life (Brown 2015). As a corollary, it is also a programmatic challenge to all collective obstacles to capital accumulation and a system of subjectivation producing possessive individuals who identify themselves solely, "as market agents. . . . and autonomous, self-interested individuals" (Binkley 2009; Bourdieu 1998). In 2010, archaeologist Chris Matthews cited the institution of subjectivity as a major effect of capitalism's emergence as a structuring agent of everyday life. This subjectivity produces "distinct individuals" who come to believe they freely choose the nature of the social relations they enter. Understood in opposition to co-constituted relations of kinship, community, ethnicity, and so on, these formations produce the contradictory position of individuals "manufactured . . . from within and against their original communities" (Matthews 2010: 11).

I propose that Kafka's parable "Before the Law," introduced in Chapter 1, is an index to understanding the history of the subject in the twentieth century. In my reading, Kafka's parable reflects the manner in which the desires of early–twentieth century subjects have been exactingly interpellated within the political economy of the present age. The individualized fate of the man from the country, before the gate constructed specifically to block him, is an ironic perversion of his belief that "the Law" be available to everyone and at all times. The significance of the parable and its deeper relevance to this work becomes clear when we conceive of the Law in the parable as not corresponding to the historically produced juridical apparatus of legal pronouncements, courts, and penal apparatuses, but rather what psychoanalyst Jacques Lacan describes as "the big Other" (Lacan 1955; Žižek 1989: 42–47).

The big Other of the Law corresponds to the unseen forces of mass signification structuring the logic of society. The Law in this context can be roughly understood as analogical to the structuring of language, while not precisely consisting of language. Operating in opposition to individual desires and experiences, the big Other provides them with words, meanings, and forms for experience and action. Lacan reminds us that we cannot speak without using the words and meanings of others, and this is the same for the formal aspects of our experience of desire, belonging, and satisfaction (Lacan 1955). Through freely accepting their membership into the Law, individuals are disciplined through the structure and form given to their desires via language and signification: an operation independent of the actual granting of those desires. Žižek suggests that ideology exists in the intimate relationship between individual choices and desires and the promises the big Other offers us:

> Every belonging to a society involves a paradoxical point at which the subject is ordered to embrace freely, as the result of his choice, what is anyway imposed on him (we must all love our country, our parents . . .). This paradox of willing (choosing freely) what is in any case necessary, of pretending (maintaining the appearance) that there is a free choice although in fact there isn't, is strictly codependent with the notion of an empty symbolic gesture, a gesture—an offer—which is meant to be rejected. (Žižek 1997: 36)

In Kafka's parable, the man believes that he has freely chosen to approach and enter the gate of the Law. It is, in fact, already open and can therefore be entered by the man at any moment. And yet, something else is peculiar about the revelation that comes at the end of the parable. The man comes believing that the gate is an objective structure he can simply enter, an impersonal fact he approaches from without.

Agamben addresses the paradoxes in Kafka's parable by proposing that the Law needs nothing to exist other than to be open and to have the excluded presence of the man to exist. Of the petitioner to the Law, he writes that "the open door destined only for him includes him in excluding him and excludes him in including him. And this is precisely the summit and the root of every law" (Agamben 1998: 50). With the suggestion of the individualized nature of the Law revealed at the parable's conclusion, we can conclude that the Law is, in a sense, always already tangled up in the materialization of the desires and beliefs of the petitioner. It has been made only for him, both in the manner that compels him to enter it and learn its secrets, and the manner in which it excludes him from ever entering. In the belief of the man from the country that the Law is "accessible to each of us at all times" and in the nature of bribes made as mere symbolic sacrifices, the foundation of the Law is revealed to be an apparatus of satisfaction and pacification, but one that owes its existence to our presence exterior to it.

This situation reflects the manner in which the history of migration is entangled with the development of the boundaries of sovereign states as discussed in Chapters 1 and 4. In the introductory chapter, I suggested that border control policies of the early twentieth century laid the ideological groundwork for the individuation of all subjects, and that the rise of neoliberalism and governmentality in the latter end of the century is its culmination. Scholars of immigration have suggested that the development of individual passports came not only from a governmental effort to identify favored and unfavored citizens, but also to separate the individual from the collectivity, which presented the threat of collectivized combination among the nation's immigrant labor force. The essential unit of movement in this equation was the individual, separated from collectivities of kinship, informal social networks, labor contracting, or other institutions that frequently accompanied past immigration (McKeown 2008: 11).

Once collectives could be individuated into subjects, the potential for introducing new forms of desire and fantasy could be accomplished. In Chapter 5, I discussed the role of psychoanalysis in the formulation of machinic mass consumerism. In this era, a coalition of elites, academics, businessmen, government planners, advertisers, and product designers applied the findings of mass psychology and propaganda techniques to create a world of endless satisfaction coupled to endless desire. Populist pursuit of the material abundance issuing out of Machine Age industry was exactingly mirrored in the phantasmagoria of department stores, catalogs, advertisements, and media presentations. Individual consumer desire was calculated to be unsatisfiable, as the motivation for mass consumerism was not redistribution but the desire to turn surplus production into surplus profit. In contrast to the disciplining of subjects in previous eras, mass consumerism of this era consists of a controlled *undisciplining* of subjects, a channeling, sublimating, or unleashing of individuated desires operationalized in material form.

Throughout the period of the late twentieth century, collectively defined subjectivities such as those defined by family, ethnicity, nation, and class were slowly weeded out of distribution by neoliberalism

(Bourdieu 1998). The only alternative template for contemporary subjectivity, suggests Foucault, is that of the entrepreneur. For the majority of the population, however, to be an entrepreneur means something far different from being free to invest one's capital in subjectively creative means; rather, it means "being compelled to manage declining wages and income, precarity, unemployment, and poverty in the same way one would manage a corporate balance-sheet" (Lazzarato 2014: 9). Lazzarato's figure of "the indebted man" is the dark side of the entrepreneur, one who must take on the responsibility and guilt for his lot to make up for the "economic, social and political failures of the neoliberal power bloc, exactly those failures externalized by the State and business onto society" (2012, 2014: 10).

In Chapter 7, I asserted that the urban renewal movement resulted not only in drastic changes in the material and economic landscape, but was also a process of subjectivization. Governmentality, as I describe it here, installs negative freedom as the defining logic of liberty in late modernity, particularly in cases in which an individual right to things such as privacy, profit, or privilege are already considered integral to broader political or economic ideologies. Negative freedom, defined as an individual's right to follow one's impulses, is enunciated only as freedom from the constraints of the other (Schroeder and Carlson 2000: 660–661). Increasingly, residents subjected each other to bureaucratic demands suggesting that, at least tactically, they had adopted the language of renewal and management as understood within the framework of negative freedom. Condition controls structured as negative freedoms unambiguously positioned the legal rights of individuals over traditional or unwritten community values. As such, they enunciated a new community comprised of atomized individuals, adopting entrepreneurial attitudes to space, labor, governance, and accumulation. Material aspects of the space, which once amounted to so much more than mere aesthetics in previous eras, were recoded as nuisances and removed. At stake was the capacity for the materiality of landscape to remember, reproduce, and channel social relations in a manner responsive to the exigencies of uncertain economy.

Negative freedom "ends up being all form and no content. When negative freedom obtains a content, it is supplied by our inclination—our pathological side, the side of feeling. In the end, mere negative freedom is, in fact, slavery to impulse" (Schroeder and Carlson 2000: 660). The compulsion to be free proposed by negative freedom can accept no content other than that which is arbitrary or contingent. In such a context, the Law, which represents the limits of freedom, bases these limits entirely upon the contingency produced by our impulses and the pure arbitrariness of individual pathology. In the context of negative freedom, entrance to the Law is impossible.

Arendt (1958) argues that a mass society composed of atomized individuals was the foundation for the totalitarian politics of the mid–twentieth century. Purged of the segmentation of local interests such as class, family, and church, atomized individuals are opened up to individualized identification and synchronization with the totalitarian state (Toscano 2017). Deleuze and Guattari (1987: 228) write poetically of the dangers of energizing the "microfascisms" of mass society:

> A multitude of black holes may very well not become centralized, and acts instead as viruses adapting to the most varied situations, sinking voids in molecular perception and semiotics. Interactions without resonance. Instead of the great paranoid fear, we are trapped in a thousand little monomanias, self-evident truths, and clarities that gush from every black hole and no longer form a system, but are only rumble and buzz, blinding lights giving everybody the mission of self-appointed judge, dispenser of justice, policeman, neighbourhood SS man.

The right-leaning populist movements identified by Kazin (1995) have erupted periodically throughout the end of the twentieth and early twenty-first centuries, most recently energized by the coordinated spread of information by mass social media—echoing the analysis of radio's "standardizing effects" by Cantrill and Allport (1935) more than eighty years ago.

In a last interview given before his mysterious death in 1975, Italian filmmaker Pier Paolo Pasolini (1975) reflected on the transformations wrought by late capitalism in Italy: "I consider consumerism a worse fascism than the classical one, because [it] did not transform Italians. It did not get into them. It was totalitarian but not totalizing." If the observed link between fascism and consumerism Pasolini makes is to be taken seriously, the stakes are high. For a critical archaeology that endeavors to apprehend and respond to the material realities of the present, developing a nuanced materialist periodicity of the latter half of the twentieth century is a mandatory step that must precede any endeavor to historicize the present (González-Ruibal 2008; Rathje et al. 2001).

THE DARKNESS AND LIGHTNESS OF LABOR AND IMMIGRATION HERITAGE

As noted by Paul Shackel, although the story of American industrialism is well documented in historic sites of national memory, the story of labor has often been left out, forgotten, or suppressed (Shackel 2004, 2009; Smith et al. 2011). This is doubly the case with the stories of immigrant laborers, who have a long history of being at times both appropriated and restricted by shifting national boundaries. Such collective memory loss is more than simply a symptomatic aspect of our accelerating information sharing within Late Capitalism. Rather, forgetting is central to the maintenance of political ideologies and the stultification of the public's capacity to imagine alternatives (Connerton 2009; Giroux 2013; González-Ruibal 2008; Huyssen 2003).

Some scholars suggest that, at present, the political economy that influences the narrative constitution of social memory has put our memory-making capacity at risk. They point to the growing corporate consolidation of education, politics, media, and other forms of industry (Chomsky 1989; Giroux 2013; Herman and Chomsky 2002). Today, this hegemony threatens with the power of what Henry Giroux (2013) has termed the "disimagination machine." Through the "violence of organized forgetting," elites have gained the power to define the ideological givens underlying public debate. Giroux warns:

> America has become amnesiac—a country in which forms of historical, political, and moral forgetting are not only willfully practiced but celebrated. The United States has degenerated into a social order that . . . views critical thought as both a liability and a threat. . . . Politicians . . . are symptomatic of a much more disturbing assault on critical thought, if not rational thinking itself. (Giroux 2013)

Neglecting to actively maintain the collective memory of immigrant labor history in the present produces a crucial aporia in the national narrative of the twentieth century.

In the early part of the twentieth century, Walter Benjamin founded a social theory of modernity on the catastrophe of progress, which produced "wreckage upon wreckage," a "pile of debris . . . growing skyward" (Benjamin 1970: 259–260). Archaeology conducted in the anthracite region reveals an alternative narrative of the twentieth century, which juxtaposes bullets with consumer waste. Deprived of the obscuring polemics of grand progressive narratives, archaeological history is rich in microscopic detail. Told through ruins, surpluses, and waste, it resolves contradictions into truths that can be difficult to swallow.

It also reveals hidden moments of utopian striving, revolutionary flashes, communal intimacy, and emancipation. The strike that preceded the Lattimer Massacre was one such moment, when a diverse workforce from Eastern and Southern Europe, doubly oppressed as migrant labor, came together to challenge a hegemonic force with their collective strength. When the community took control of the coal company briefly in the 1930s, they overturned decades of vertical control, righting a bankrupt system. At the time, communal systems of interdependence, decision making, and social solidarity were flourishing. In the years that followed, residents and outsiders evoked the memory of the Massacre to inspire class solidarity among workers of all backgrounds.

In 1972 Cesar Chavez attended the ceremony to commemorate the monument to the Lattimer Massacre. It was the 75th anniversary of the event. He was busy at the time advocating for a boycott of lettuce to protect the mostly Mexican workers of the lettuce-growing industry (*New York Times*, September 4, 1972: 4). Although the speech he delivered was not recorded, his notes are preserved in the archives of the United Farm Workers. Chavez drew a direct connection between the immigrant marchers at Lattimer and the Mexican migrant workers. A schematic from his speech emerges from his notes:

4) In their 75th year—gather at site where good men
Peaceful Union Men—
Seeking bread + Justice—were massacred

in an unprecedented Act of Brutality
—endowing all working men + women
—Inheritance—Generation

5) No Better treatment to Power of Non-Violence +
Might of Sacrifice to be
Found anywhere in this land
6) These Martyred Immigrants
—only crime at the time
their poverty—their strange sounding names—
their foreign tongues.
—If could speak to us today
They would tell us plainly
—in the words of another labor martyr
—DONT MOURN FOR US, ORGANIZE.
7) It [is] fitting + proper, unveil memorial plaque commemorating the 75th year at
Tragic Event
—For only in Preserving Labor's Past, can we preserve labor's future.

NOTES

Chapter 1. An Archaeology of the Twentieth Century in the Anthracite Coal Region of Northeast Pennsylvania

1. Contemporary narratives of the Massacre and its social context have been written by a few historians synthesizing accounts collected in newspapers, official documents, and the scraps of trial transcripts that have survived therein. The first piece of literature devoted wholly to the Massacre is Edward Pinkowski's 1950 *The Lattimer Massacre*. Accounts by Victor Greene and George Turner followed, providing contextual analysis derived from their intellectual roots in new social and labor histories (Greene 1964, 1968; Turner 1977). The most recent, best known, and only popular account of the event comes from Michael Novak's 1978 historical fiction *The Guns of Lattimer*, reprinted in 1996. A number of essays followed providing analysis and context (Aurand 2002; Beik 2002; Blatz 2002; Culen 1977; Dubofsky 2002; Miller and Sharpless 1998; Stolarik 2002; Turner 1984; Wolensky and Hastie 2013).

Chapter 3. The Memory of Violence: The Lattimer Massacre of 1897

1. In fact, census and company records indicate this was not entirely the case, at least by the turn of the century, although it may have been true for the period leading up to the Massacre, as the missing 1890 census is not present to provide us with evidence supporting this assertion.

2. In reviewing the history of past strikes in the region including those of 1877, 1885, and 1888–1899, scholars have found no grounds to indicate new immigrants are to blame for the failure of organization in the region. Rather, they find the failure to propose effective union policy, the inability to coordinate regional strike efforts, and united operator resistance to explain the weakness of labor in these fights (Blatz 1994; Greene 1964, 1968; Wolensky and Hastie 2013). Importantly, by taking this evasive and scapegoating solution, historians sidelined the causal factors of operator manipulation of the labor market, as well as the effects

of de-skilling and mechanizing production processes. Each of these practices disempowered labor by segmenting solidarities and communities.

3. Last names of posse members noted during the subsequent trial included: Anderson, Babcock, Barton, Cook, Diehl, Hess, McShea, Mulhall, Nichols, Osborne, Pardee, Platt, Siewell, Turnbach, Warriner, and Zierdt.

4. Jeff Speakmen, Head of Technical Studies at the Smithsonian Museum Conservation Institute, contributed an XRF analysis of the bullets to this research.

Chapter 4. Migration, Part 1: The Social and Material Landscape of the Company Town

1. A connecting link between two larger building elements.

Chapter 5. The Archaeology of Machinic Mass Consumerism and the Logistics of the Factory Floor in Everyday Life

1. For critiques of this scholarly approach, see Graeber 2011; Lodziak 2002; Wurst and McGuire 1999.

REFERENCES

Adorno, Theodor, and Max Horkheimer
1944 *The Dialectic of Enlightenment.* Herder and Herder, New York.

Agamben, Giorgio
1998 *Homo Sacer: Sovereign Power and Bare Life.* Stanford University Press, Stanford, California.

Aldrich, Mark
1997 The Perils of Mining Anthracite: Regulation, Technology and Safety, 1870–1945. *Pennsylvania History* 64(3): 361–383.

Angelil, Marc, and Cary Siress
2011 Discounting Territory: Logistics as Capital Principle of Spatial Practices. In *New Geographies 4, Scales of the Earth,* Harvard Graduate School of Design, Harvard University, Cambridge, Massachusetts.

Applebaum, Binyamin
2016 In City Built by Immigrants, Immigration Is the Defining Political Issue. *N. Y. Times,* October 12, 2016. Website accessed April 2017.

Arendt, Hannah
1958 *The Origins of Totalitarianism.* World, Cleveland, Ohio.
1994 [1943] We Refugees. In *Altogether Elsewhere: Writers on Exile,* edited by Marc Robinson, pp. 110–119. Faber and Faber, Boston.

Arnold, Andrew
2014 *Fueling the Gilded Age.* New York University Press, New York.

Aurand, Harold
1968 The Anthracite Strike of 1887–1888. *Pennsylvania History* 35(2): 169–185.
1970 Diversifying the Economy of the Anthracite Regions, 1880–1900. *Pennsylvania Magazine of History and Biography* 94(1): 54–61.
1977 Social Motivation of the Anthracite Workers: 1901–1920. *Labor History* 18(3): 360–365.

1985 Mine Safety and Social Control in the Anthracite Industry. *Pennsylvania History* 52(4): 227–241.

1991 Early Mine Workers' Organizations in the Anthracite Region. *Pennsylvania History* 58(4): 298–310.

2002 The Lattimer Massacre: Who Owns History?—An Introduction. *Pennsylvania History* 69(1): 5–10.

2003 *Coalcracker Culture: Work and Values in Pennsylvania Anthracite, 1835–1935*. Susquehanna University Press, Selinsgrove, Pennsylvania.

Barendse, Michael

1981 Social Expectations and Perception: The case of the Slavic anthracite workers. University of Pennsylvania Press, University Park, Pennsylvania.

Barrett, James R., and David Roediger

1997 Inbetween Peoples: Race, Nationality and the "New Immigrant" Working Class. *Journal of American Ethnic History* 16(3): 3–44.

Barthes, Roland

1988 Plastic. *Perspecta* 24: 92–93.

Baylor, Chas Gano

1898 Imperialism. *Lucifer, The Light Bearer*. February 23: 57–58. Chicago, Illinois.

Beik, Mildred A.

2002 The Significance of the Lattimer Massacre: Who Owns History? *Pennsylvania History* 69(1): 58–70.

Benjamin, Walter

1970 *Illuminations*. Schocken Books, New York.

1975 Eduard Fuchs: Collector and Historian. *New German Critique* 5: 27–58.

1978 [1931] The Destructive Character. In *Reflections*. Harcourt, Brace, Jovanovich, New York.

1999 *The Arcades Project*. Belknap Press of Harvard University, Cambridge, Massachusetts.

Bernays, Edward

1923 *Crystallizing Public Opinion*. Liveright, New York.

1928 *Propaganda*. Electronic document, http://asset-4.soup.io/asset/0324/0984_4ce7.pdf, accessed September 9, 2012.

Berthoff, Rowland

1965 The Social Order of the Anthracite Region, 1825–1902. *Pennsylvania Magazine of History and Biography* 89(3): 261–291.

Bertolluci, Bernardo

1975 Films are Animal Events: Bernardo Bertolucci Talks about His New Film. Interview by Gideon Bachmann. *Film Quarterly* 29(1): 11–19.

Binkley, Sam

2009 The Work of Neoliberal Governmentality: Temporality and Ethical Substance in the Tale of Two Dads. *Foucault Studies* 6: 60–78.

Blatz, Perry K.

1994 *Democratic Miners: Work and Labor Relations in the Anthracite Coal Industry, 1875–1925*. State University of New York Press, Albany.

2002 Reflections on Lattimer: A Complex and Significant Event. *Pennsylvania History* 69(1): 42–51.

Bluestone, Barry, and Bennett Harrison

1982 *The Deindustrialization of America*. Harper Collins, New York.

Bodnar, John

1992 *Remaking America: Public Memory, Commemoration, and Patriotism in the Twentieth Century*. Princeton University, Princeton, New Jersey.

Bourdieu, Pierre

1998 *Acts of Resistance: Against the New Myths of Our Times*. Polity Press, London.

Brooks, John G.

1898 An Impression of the Anthracite Coal Troubles. *Yale Review: A Quarterly Journal for the Scientific Discussion of Economic, Political and Social Questions* 6(3): 306–311.

Brown, Wendy

2015 *Undoing the Demos*. Zone Books, New York.

Buck-Morss, Susan

1989 *The Dialectics of Seeing*. MIT Press, Cambridge.

1992 Aesthetics and Anaesthetics: Walter Benjamin's Artwork Essay Reconsidered. *October* 62: 3–41.

Burgoyne, Robert

1989 Temporality as Historical Argument in Bertolucci's "1900." *Cinema Journal* 28(3): 57–68.

Butler, Judith

2004 *Precarious Life: The Powers of Mourning and Violence*. Verso, New York.

Calavita, Kitty

1994 U.S. Immigration Policy: Contradictions and Projections for the Future. *Indiana Journal of Global Legal Studies* 2(1): 143–152.

Camp, Stacey Lynn

2011 Consuming Citizenship? The Archaeology of Mexican Immigrant Ambivalence in Early Twentieth Century Los Angeles. *International Journal of Historical Archaeology* 15: 305–328.

CAN DO

1974 *The CAN DO Story*. CAN DO Corporation, Hazleton, Pennsylvania.

Cantrill, Hadley, and Gordon Allport
1935 *The Psychology of Radio.* Harper and Brothers, New York.
Chandler, Jr., Alfred D.
1972 Anthracite Coal and the Beginnings of the Industrial Revolution in the United States. *Business History Review* 46(2): 141–181.
Chavez, Leo
2008 *The Latino Threat: Constructing Immigrants, Citizens, and the Nation.* Stanford University Press, Stanford, California.
Chomsky, Noam
1989 *Necessary Illusions: Thought Control in Democratic Societies.* South End Press, New York.
Coal Age
1917 *Coal & Coke News* 11(20): 876–877.
Cohen, Lizbeth
1986 Embellishing a Life of Labor: An Interpretation of the Material Culture of American Working-Class Homes, 1885–1915. In *Common Places, Readings in American Vernacular Architecture,* edited by Dell Upton and John Michael Vlach, pp. 261–280. University of Georgia Press, Athens, Georgia.
1990 *Making a New Deal.* Cambridge University Press, Cambridge, Massachusetts.
1996 From Town Center to Shopping Center: The Reconfiguration of Community Marketplaces in Postwar America. *American Historical Review* 101(4): 1050–1081.
2003 *A Consumer's Republic.* Alfred A. Knopf, New York.
Connerton, Paul
2009 *How Modernity Forgets.* Cambridge University Press, Cambridge.
Coutin, Susan B.
1995 Smugglers or Samaritans in Tucson, Arizona: Producing and Contesting Legal Truth. *American Ethnologist* 22(3): 549–71.
Cowie, Jefferson
1999 *Capital Moves: RCA's Seventy Year Quest for Cheap Labor.* Cornell University Press, New York.
Cronan, William
1991 *Nature's Metropolis.* W. W. Norton, New York.
Culen, Konstantin
1977 The Lattimer Massacre. *Slovakia* 27(50): 44–61.
Curtis, Adam
2002 *The Century of the Self.* Documentary. BBC Films.
Dawdy, Shannon Lee
2010 Clockpunk Anthropology and the Ruins of Modernity. *Current Anthropology* 51(6): 761–793.

de Boever, Arne

2009 Agamben and Marx: Sovereignty, Governmentality, Economy. *Law Critique* 20: 259–270.

de Certeau, Michel

1984 *The Practice of Everyday Life*. University of California Press, Berkeley.

De Cunzo, Lu Ann, and Julie H. Ernstein

2006 Landscapes, Ideology, and Experience in Historical Archaeology. In *The Cambridge Companion to Historical Archaeology*, edited by Dan Hicks and Mary C. Beaudry, pp. 255–272. Cambridge University Press, Cambridge.

Deasy, George F., and Phyllis R. Griess

1961 Tourism for the Anthracite Region—An Alternative for Unemployment. *Mineral Industries* (30)7: 1–8.

1965 Effects of a Declining Mining Economy on the Pennsylvania Anthracite Region. *Annals of the Association of American Geographers* 55(2): 239–259.

Deleuze, Gilles, and Felix Guattari

1987 *A Thousand Plateaus*. University of Minnesota Press, Minneapolis.

2009 *Anti-Oedipus: Capitalism and Schizophrenia*. Penguin Books, New York.

Dick, Hilary Parsons, and Kristina Wirtz

2011 Introduction—Racializing Discourses. *Journal of Linguistic Anthropology* 21(1): 2–10.

Dillingham Commission (United States Immigration Commission)

1911a Reports of the Immigration Commission, Vol. 16, Immigrants in Industries: Anthracite Coal Mining. Government Printing Office, Washington, D.C.

1911b Reports of the Immigration Commission, Vol. 6, Immigrants in Industries: Bituminous Coal Mining. Government Printing Office, Washington, D.C.

1911c Reports of the Immigration Commission, Vol. 5, Dictionary of Races and Peoples. Government Printing Office, Washington, D.C.

Dinius, Oliver J., and Angela Vergara (editors)

2011 *Company Towns in the Americas: Landscape, Power, and Working-Class Communities*. Georgia University Press, Athens.

Dublin, Thomas, and Walter Licht

2005 *The Face of Decline: The Pennsylvania Anthracite Region in the Twentieth Century*. Cornell University Press, Ithaca, New York.

Dubofsky, Melvyn

1991 The Origins of the Labor Movement in the United States: Themes from the Nineteenth Century. *Pennsylvania History* 58(4): 269–277.

1996 *Industrialism and the American Worker, 1865–1920*. Harlan Davidson, Wheeling, Illinois.

2002 The Lattimer Massacre and the Meaning of Citizenship. *Pennsylvania History* 69(1): 52–57.

Engels, Friedrich
1969 *The Condition of the Working Class in England.* Academy, Chicago, Illinois.

Epperson, Terrance W.
2000 Panoptic Plantations: The Garden Sights of Thomas Jefferson and George Mason. In *Lines That Divide: Historical Archaeology of Race, Class, and Gender.* University of Tennessee Press, Nashville.

Ewen, Stuart
1976 *Captains of Consciousness.* McGraw-Hill, New York.
1988 *All Consuming Images: The Politics of Style in Contemporary Culture.* Basic Books, New York.
1996 *PR!: A Social History of Spin.* Basic Books, New York.

Fairbanks, Robert B.
2012 The Housing Act of 1954 and the War against Slums in the Southwestern United States. Paper presented at the 15th International Planning History Society Conference, Sao Paulo, Brazil.
2014 *The War on Slums in the Southwest: Public Housing and Slum Clearance in Texas, Arizona, and New Mexico, 1935–1965.* Temple University Press, Philadelphia, Pennsylvania.

Falk, Candace (editor)
2003 *Emma Goldman: A Documentary History of the American Years: Made in America, 1890–1901, Vol. 1.* University of California Press, Berkeley.

Fassin, Didier
2011 Policing Borders, Producing Boundaries: the Governmentality of Immigration in Dark Times. *Annual Review of Anthropology* 40: 213–26.

Filene, Edward A.
1929 Mass Production Makes a Better World. *Atlantic Monthly* 143: 625–631.
1930 The New Capitalism. *Annals of the American Academy of Political and Social Science, Part 1: The Second Industrial Revolution and Its Significance* 149: 3–11.
1934 *The Consumer's Dollar.* John Day Pamphlets, No. 41, New York.

Fisfis, Nick S., and Harold Greenberg
1962 Suburban Renewal in Pennsylvania. *University of Pennsylvania Law Review* 111(1): 61–110.

Fleury-Steiner, Benjamin, and Jamie Longazel
2010 Neoliberalism, Community Development, and Anti-Immigrant Backlash in Hazleton, Pennsylvania. In *Taking Local Control: Immigration Policy Activism in U.S. Cities and States,* edited by M. Varsanyi, pp. 157–172. Stanford University Press, Stanford, California.

Foucault, Michel
1984 Of Other Spaces, Heterotopias. *Architecture, Mouvement, Continuité* 5: 46–49.
2009 *Security, Territory, Population: Lectures at the Collège de France 1977–1978.* Picador Press, New York.
Foulke, Calvin Pardee, and William G. Foulke
1979 *Calvin Pardee, 1941–1923: His Family and His Enterprises.* Pardee Company, Philadelphia, Pennsylvania.
Francaviglia, Richard
1991. *Hard Places: Reading the Landscapes of America's Mining Districts.* Johns Hopkins University Press, Baltimore.
Gabaccia, Donna R., and Fraser M. Ottanelli
2001 *Italian Workers of the World: Labor Migration and the Formation of Multiethnic States.* University of Illinois Press, Urbana.
Giroux, Henry
2013 *The Violence of Organized Forgetting.* Electronic document, http://www.truth-out.org/opinion/item/17647-the-violence-of-organized-forgetting, accessed May 16, 2014.
Gobin, J.P.S.
1898 Reports Relating to the Tour of Duty of the Third Brigade, N.G.P. and First Troop P.C.C. 1st. Brigade, NGP at Hazleton Luzerne County Pennsylvania, Sept. 11 to Oct. 4 1897. Annual Report of the Adjutant General of Pennsylvania, Transmitted to the Governor of Law. For the Year 1897. State Printer of Pennsylvania, Harrisburg.
Goddard, Richard
2002 Nothing but Tar Paper Shacks. *Historical Archaeology* 36(3): 85–93.
González Ruibal, Alfredo
2008 Time to Destroy: An Archaeology of Supermodernity. *Current Anthropology* 49(2): 247–279.
González-Tennant, Edward
2018 *The Rosewood Massacre: An Archaeology and History of Intersectional Violence.* University Press of Florida, Gainesville, Florida.
Graeber, David
2004 *Fragments of an Anarchist Anthropology.* Prickly Paradigm, Chicago.
2011 Consumption. *Current Anthropology* 52(4): 489–511.
Gramsci, Antonio
1971 *Selections from the Prison Notebooks,* translated and edited by Quintin Hoare and Geoffrey Nowell Smith. International, New York.
Green, Hardy
2010 *The Company Town: The Industrial Edens and Satanic Mills That Shaped the American Economy.* Basic Books, New York.
Greene, Julie
2009 *The Canal Builders.* Penguin Books, New York, NY.

Greene, Victor R

1964 A Study in Slavs, Strikes, and Unions: The Anthracite Strike of 1897. *Pennsylvania History* 31(2): 199–215.

1965 The Poles and Anthracite Unions in Pennsylvania. *Polish American Studies* 22(1): 10–16.

1968 *The Slavic Community on Strike: Immigrant Labor in Pennsylvania Anthracite*. University of Notre Dame Press, Notre Dame, Indiana.

Guattari, Felix

2009 *Chaosophy*. Semiotext(e), Pasadena, California.

2011 *The Machinic Unconscious*. Semiotext(e), Pasadena, California.

Guglielmo, Thomas A.

2003 *White on Arrival: Italians, Race, Color, and Power in Chicago, 1890–1945*. Oxford University Press, New York.

Gutman, Herbert

1977 *Work, Culture and Society in Industrializing America*. Vintage Books, New York.

Hambidge, Jay

1898 An Artist's Impression of the Colliery Region. *Century* 55(6): 822–828.

Hanchett, Thomas W.

1996 U.S. Tax Policy and the Shopping-Center Boom of the 1950s and 1960s. *American Historical Review* 101(4): 1082–1110.

Haraway, Donna

1990 A Manifesto for Cyborgs: Science, Technology, and Socialist Feminism in the 1980s. In *Feminism/Postmodernism*, edited by Linda J. Nicholson, Routledge, New York.

Hardesty, Donald L.

1998 Power and the mining community in the American West. In *Social Approaches to an Industrial Past: the Archaeology and Anthropology of Mining*, edited by A. Bernard Knapp, Vincent Pigott, Eugenia Herbert. Routledge, London.

Harvard Business School

2014 National Union Radio Corporation. Lehman Brothers Archives—Contemporary Business Archives. Baker Library Historical Collections. Electronic document, http://www.library.hbs.edu/hc/lehman/company.html?company=national_union_radio_corp, accessed December 15, 2014.

Harvey, David

1985 *The Urbanization of Capital*. Basil Blackwell, Oxford.

1989 *The Condition of Postmodernity: An Enquiry into the Origins of Cultural Change*. New York: Wiley.

1990 Flexible Accumulation through Urbanization: Reflections on "Post-Modernism" in the American City. *Perspecta* 26: 251–272.

2001 The Geography of Capitalist Accumulation. In *Spaces of Capital Towards a Critical Geography*, pp. 237–266. Routledge, New York.

2005 *A Brief History of Neoliberalism.* Oxford University Press, Oxford.

2011 *The Enigma of Capital.* Oxford University Press, Oxford.

2014 *Seventeen Contradictions and the End of Capitalism.* Profile Books, London.

Hatcher, Julian S.

1966 *Hatcher's Notebook.* Stackpole Company, Harrisburg, Pennsylvania.

Herman, Edward S., and Noam Chomsky

2002 *Manufacturing Consent: The Political Economy of the Mass Media.* Pantheon Books, New York.

Herod, Andrew

2011 Social Engineering through Spatial Engineering. In *Company Towns in the Americas: Landscape, Power, and Working-Class Communities*, edited by Oliver J. Dinius and Angela Vergara, pp. 21–40. Georgia University Press, Athens.

Hickel, Jason, and Arsalan Khan

2012 The Culture of Capitalism and the Crisis of Critique. *Anthropological Quarterly* 85(1): 203–227.

Hobsbawm Eric J.

1959 *Primitive Rebels.* University of Manchester Press, Manchester, UK.

1994 *The Age of Extremes: The Short Twentieth Century, 1914–1991.* Vintage Books, New York.

Hodder, Ian

2012 *Entangled.* Wily Blackwell, Malden, Massachusetts.

Hoerder, Dirk

2002 *Cultures in Contact: World Migrations in the Second Millennium.* Duke University Press, Durham, North Carolina.

Hoyt, Henry

1898 Memorandum to Department of State on Facts In Commonwealth vs. James L. Martin, Sheriff et al. Court of Oyer and Terminer, Luzerne County, Penn'a, February–March, 1898. In Papers Relating to the Foreign Relations of the United States, with the Annual Message of the President Transmitted to Congress December 5, 1898, pp. 97–109. Government Printing Office, Washington D.C.

Huyssen, Andreas

2003 *Present Pasts: Urban Palimpsests and the Politics of Memory.* Stanford University Press, Stanford, California.

Inspectors of the Coal Anthracite Region (ICMAR)

1870–1938 *Reports of the Inspectors of Coal Mines of the Anthracite and Bituminous Coal Regions of Pennsylvania.* Reports on file, Archive.org.

Itter, William A.

1934 Early Labor Troubles n the Schuylkill Anthracite District. *Pennsylvania History* 1(1): 28–37.

Jackson, Kenneth T.
1996 All the World's a Mall: Reflections on the Social and Economic Consequences of the American Shopping Center. *American Historical Review* 101(4): 1111–1121.
Jameson, Fredric
2009 *Valences of the Dialectic*. Verso, London.
2011 *Representing Capital*. Verso, London.
Jerome, Harry
1934 *Mechanization in Industry*. National Bureau of Economic Research, New York.
Johnson, Bob
2010 "An Upthrust into Barbarism": Coal, Trauma, and Origins of the Modern Self, 1885–1951. *Journal of American Culture* 33(4): 265–279.
Kafka, Franz
1971 Before the Law. In *The Complete Stories*, edited by Nahum M. Glatzer. Schocken Books, New York.
Kazin, Michael
1995 *The Populist Persuasion*. Basic Books, NY.
Keil, Thomas J., and Wayne M. Usui
1992 Split Labor Markets in the 19th Century Anthracite Coal Fields. *Polish American Studies* 49(2): 19–42.
Kendree and Shepherd
1967a Hazle Township Comprehensive Plan: Part One, Development Patterns. Submitted to Hazle Township Planning Commission by Kendree and Shepherd, Planning Consultants. Electronic document, http://elibrary.pacounties.org/Documents/Luzerne_County/1592;%20Hazle%20Township/4207933376mcp.pdf, accessed December 10, 2014.
1967b Hazle Township Comprehensive Plan: Part Two, Development Proposals. Submitted to Hazle Township Planning Commission by Kendree and Shepherd, Planning Consultants. Electronic document, http://elibrary.pacounties.org/Documents/Luzerne_County/1592;%20Hazle%20Township/4207933376mcp.pdf, accessed December 10, 2014.
1967c Hazle Township Comprehensive Plan: Part Three, Development Procedures. Submitted to Hazle Township Planning Commission by Kendree and Shepherd, Planning Consultants. Electronic document, http://elibrary.pacounties.org/Documents/Luzerne_County/1592;%20Hazle%20Township/4207933376mcp.pdf, accessed December 10, 2014.
Kobialka, Matthew D.
2012 Immigration and Language Threat in a Local Newspaper: A Case

Study of Hazleton, Pennsylvania. Master's thesis, Department of Education, University Of Puerto Rico, Mayagüez Campus.

Kramer, Paul A.

2006 *The Blood of Government: Race, Empire, the United States, and the Philippines.* University of North Carolina Press, Chapel Hill.

Kristeva, Julia

1991 *Strangers to Ourselves.* Columbia University Press, New York.

Lacan, Jacques

1955 *The Seminar of Jacques Lacan: Book II, The Ego in Freud's Theory and in the Technique of Psychoanalysis, 1954–1955,* edited by Jacques-Alain Miller, translated by Sylvana Tomaselli. W. W. Norton, New York.

Lattimer Coal Company Records (LCC)

1900–1945 Private Collection of Joseph Michel. Hazleton, Pennsylvania.

Lattimer Massacre Project (LMP)

2012 RE: Employee Record Cards—Public Comments. Electronic document, http://lattimermassacre.wordpress.com/2011/12/21/employee-record-cards-part-ii/, accessed January 15, 2012.

Lazzarato, Maurizio

2012 *The Making of the Indebted Man: An Essay on the Neoliberal Condition,* translated by Joshua David Jordan. Semiotext(e), Cambridge, Massachusetts.

2014 *Signs and Machines: Capitalism and the Production of Subjectivity,* translated by Joshua David Jordan. Semiotext(e), Cambridge, Massachusetts.

Leach, Eugene E.

1986 Mastering the Crowd: Collective Behavior and Mass Society in American Social Thought, 1917–1939. *American Studies* 27(1): 99–114.

1992 "Mental Epidemics": Crowd Psychology and American Culture, 1890–1940. *American Studies* 33(1): 5–29.

Leach, Neil

2005 9/11. In *Urban Memory,* edited by Mark Crinson, pp. 169–191. Routledge, London.

Le Bon, Gustave

1895 *The Crowd: A Study of the Popular Mind.* Batoche Books, Kitchener, Ontario.

Leone, Mark P.

1995 A Historical Archaeology of Capitalism. *American Anthropologist*97(2): 251–268.

2005 *The Archaeology of Liberty in an American Capital: Excavations in Annapolis.* University of California Press, Berkeley.

2010 *Critical Historical Archaeology.* Left Coast Press, Walnut Creek, California.

Leone, Mark P., and Paul Shackel
1987 Forks, Clocks, and Power. In *Mirror and Metaphor,* edited by Daniel J. Ingersoll and Gordon R. Bronitski, pp. 45–61. University Press of America, Lanham, Maryland.
Leone, Mark P., Parker B. Potter, Jr., and Paul A. Shackel
1987 Toward a Critical Archaeology. *Current Anthropology* 28(3): 283–302.
Little, Barbara, and Paul Shackel
2014 *Archaeology, Heritage and Civic Engagement.* Left Coast Press, Walnut Creek, California.
Locomotive Firemen's Magazine (LFM)
1898 The Hazleton Slaughter. *Locomotive Firemen's Magazine* 23(5): 365–371.
Lodziak, Conrad
2002 *The Myth of Consumerism.* Pluto Press, London.
Logan, William, and Keir Reeves
2009 *Places of Pain and Shame: Dealing with "Difficult Heritage."* Routledge, London.
Longazel, Jamie
2013a Subordinating Myth: Latino/a Immigration, Crime, and Exclusion. *Sociology Compass* 7(2): 87–96.
2013b Hazleton Historical Society Community Forum: The Future of Hazleton. June 5, 2013, Penn State University, Hazleton Campus, Hazleton, Pennsylvania.
2014 Rhetorical Barriers to Mobilizing for Immigrant Rights: White Innocence and Latina/o Abstraction. *Law & Social Inquiry* 39(3): 580–600.
2016 *Undocumented Fears: Immigration and the Politics of Divide and Conquer in Hazleton, Pennsylvania.* Temple University Press, Philadelphia, Pennsylvania.
Longazel, Jamie, and Benjamin Fleury-Steiner
2011 Exploiting Borders: The Political Economy of Local Backlash against Undocumented Immigrants. *Chicana/o-Latina/o* 43: 43–64
Lukács, Georg
1971 *History and Class Consciousness.* MIT Press, Cambridge, Massachusetts.
Lyster, Clare
2012 The Logistical Figure. *Cabinet* 47: 55–62.
Marsh, Ben
1987 Continuity and Decline in the Anthracite Towns of Pennsylvania. *Annals of the Association of American Geographers* 77(3): 337–352.
McAtackney, Laura
2013 *An Archaeology of the Troubles: The Dark Heritage of Long Kesh/Maze Prison.* Oxford University Press, Oxford.

McGuire, Randall H.

2002 *A Marxist Archaeology*. Percheron Press, New York.

2008 *Archaeology as Political Action*. University of California Press, Berkeley.

McKanders, Karla M.

2007 Welcome to Hazleton ("Illegal" Immigrants Beware): Local Employment Immigration Ordinances and What the Federal Government Must Do about It. *Loyola University Chicago Law Journal*, Research Paper No. 65, University of Tennessee Legal Studies, Knoxville, Tennessee.

McKeown, Adam M.

2008 *Melancholy Order: Asian Immigration and the Globalization of Borders*. Columbia University Press, New York.

Macdonald, Sharon

2010 *Difficult Heritage: Negotiating the Nazi Past in Nuremberg and Beyond.* Routledge, London.

Marx, Karl

1972 *The Grundrisse*. Translated by David McLellan. Harper and Row, New York.

1977 *Capital: Vol. 1*. Vintage Books, New York.

1978 Grundrisse—Outlines of the Critique of Political Economy. In *The Marx-Engels Reader*, 2nd ed. W. W. Norton, New York.

2003 [1852] *The Eighteenth Brumaire of Louis Bonaparte*. Marxist.org. Electronic document, https://www.marxists.org/archive/marx/works/1852/18th-brumaire/, accessed February 6, 2015.

Matthews, Chris

2002 *An Archaeology of History and Tradition: Annapolis and the Making of the Modern Landscape*. Kluwer Academic/Plenum Press, New York.

2010 *The Archaeology of American Capitalism*. University of Florida Press, Gainesville.

Mattick, Paul

1938 What Can the Unemployed Do? *Living Marxism* (4)3: 85–92

Mbembe, Achille

2001 *On the Postcolony*. University of California Press, Berkeley.

Metheny, Karen B.

2007 *From the Miner's Doublehouse: Archaeology and Landscape in a Pennsylvania Coal Company Town*. University of Tennessee Press, Knoxville.

Miller, E. Willard

1955 The Southern Anthracite Region a Problem Area. *Economic Geography* 31(4): 331–350.

Miller, Donald, and Richard E. Sharpless

1998 *The Kingdom of Coal: Work, Enterprise, and Ethnic Communities in the*

Mine Fields. Canal History and Technology Press, Easton, Pennsylvania.

Montgomery, David
1987 *The Fall of the House of Labor: The Workplace, the State, and American Labor Activism, 1865–1925*. Cambridge University Press, Cambridge, Massachusetts.

Mowry, George E., and Blaine A. Brownell
1981 *The Urban Nation: 1920–1980*. Hill & Wang, New York.

Mrozowski, Stephan
2006 *The Archaeology of Class in the United States*. Cambridge University Press, Cambridge, Massachusetts.
2014 Imagining an Archaeology of the Future: Capitalism and Colonialism Past and Present. *International Journal of Historical Archaeology* 18: 340–360.

Mullins, Paul
1999 *Race and Affluence: An Archaeology of African American Consumer Culture*. Kluwer/Plenum, New York.

Mulrooney, Margaret
1989 *A Legacy of Coal: The Coal Company Towns of Southwestern Pennsylvania*. Historic American Buildings Survey/Historic American Engineering Record. National Park Service, Washington, D.C.

Nassaney, Michael M., and Marjorie R. Abel
2000 Urban Spaces, Labor Organization, and Social Control: Lessons from New England's Nineteenth-Century Cutlery Industry. In *Lines That Divide: Historical Archaeologies of Race, Class, and Gender*, edited by James A. Delle, Stephen A. Mrozowski, and Robert Paynter, pp. 239–275. University of Tennessee Press, Knoxville.

Ngai, Mae
1999 The Architecture of Race in American Immigration Law: A Reexamination of the Immigration Act of 1924. *Journal of American History* 86(1): 67–92.
2004a Race, Nation, and Citizenship in Late Nineteenth Century America. In *The Columbia Documentary History of Race and Ethnicity in America*, edited by Ronald H. Bayor, pp. 309–336. Columbia University Press, New York.
2004b *Impossible Subjects: Illegal Aliens and the Making of Modern America*. Princeton University Press, Princeton, New Jersey.

Novak, Michael
1996 *The Guns of Lattimer*. Transaction, New Brunswick, New Jersey.

Olsen, Bjørnar
2010 *In Defense of Things*. Altamira, Lanham, Maryland.

Oshinsky, David
1995 "The Last Refuge of Everybody: Review of Michael Kazin 'The Populist Persuasion'" *New York Times*, February 12, 1995.

Palladino, Grace
1990 *Another Civil War.* University of Illinois Press, Urbana.
Palmer, Henry
1913 *Fifty Years at the Bar and in Politics.* Snyder and Bischof, Williamsport, Pennsylvania.
Palus, Matthew A.
2010 Materialities of Government: A Historical Archaeology of Infrastructure in Annapolis and Eastport, 1865–1951. Unpublished PhD dissertation, Department of Anthropology, Columbia University, New York.
Pasolini, Pier Paolo
1975 The Lost Pasolini Interview. Webpage, https://mubi.com/notebook/posts/the-lost-pasolini-interview, accessed May 5, 2017.
Paynter, Robert
1988 Steps to an Archaeology of Capitalism. In *The Recovery of Meaning: Historical Archaeology in the United States*, edited by Mark Leone and Parker B. Potter, pp. 407–433. Smithsonian Institution Press, Washington, D.C.
Pew Research Center
2013 *Mapping the Latino Population, By State, County and City.* Pew Research Center, Washington, D.C.
Pinkowski, Edward
1950 *The Lattimer Massacre.* Sunshine, Philadelphia.
Powell, H. Benjamin
1980 The Pennsylvania Anthracite Industry, 1769–1976. *Pennsylvania History* 47(1): 3–28.
Pula, James S.
1980 American Immigration Policy and the Dillingham Commission. *Polish American Studies* 37(1): 5–31.
Rathje, William, Vincent M. LaMotta, and William A. Longacre
2001 Into The Black Hole: Archaeology 2001 and Beyond . . . In *Archaeology: The Widening Debate,* edited by Barry Cunliffe, Wendy Davies, and Colin Renfrew, pp. 497–540. British Academy, London.
Raunig, Gerald
2010 *A Thousand Machines.* Semiotext(e), Los Angeles, California.
2016 *Dividuum: Machinic Capitalism and Molecular Revolution.* Semiotext(e), South Pasadena, California.
Read, Jason
2009 A Genealogy of Homo-Economicus: Neoliberalism and the Production of Subjectivity. *Foucault Studies* 6: 25–36.
Roberts, Peter
1901 *The Anthracite Coal Industry.* Macmillan, London.
1904 *Anthracite Coal Communities.* Macmillan, London.

1905 The Sclavs in Anthracite Coal Communities. *Charities, a Review of Local and General Philanthropy* 13(10): 215–222.

Roediger, David

1999 *The Wages of Whiteness: Race and the Making of the American Working Class.* Verso, London.

2005 *Working towards Whiteness.* Basic Books, New York.

Roediger, David R., and Elizabeth D. Esch

2012 *The Production of Difference: Race and the Management of Labor in U.S. History.* Oxford University Press, New York.

Roller, Michael

2012 (Trans)national (Im)migration and Labor in the Gilded Age: The Lattimer Massacre in Perspective. Presentation at the 45th Annual Conference on Historical and Underwater Archaeology, January 4–8, 2012, Baltimore, Maryland.

2013 Rewriting Narratives of Labor Violence: A Transnational Perspective of the Lattimer Massacre. *Historical Archaeology* 47(3): 109–123.

2015 Diabolical Consumerism: Mass Psychology and Social Production between the Gilded and the Golden Ages. In *Historical Archaeologies of Capitalism*, 2nd ed., edited by Mark Leone and Jocelyn Knauf, pp. 25–50. Springer, London.

Rood, Henry E.

1898 A Pennsylvania Colliery Village: A Polyglot Community. *Century* 55(6): 809–821.

Rose, Dan

1981 *Energy Transition and the Local Community.* University of Pennsylvania Press, Philadelphia.

1989 *Patterns of American Culture.* University of Pennsylvania Press, Philadelphia.

Rose, Nikolas, and Peter Miller

1992 Political Power beyond the State: Problematics of Government. *British Journal of Sociology* 43(2): 173–205.

Ross, Edward A.

1908 *Social Psychology: An Outline and Source Book.* Macmillan, Norwood, Massachusetts.

Sayers, Daniel O.

2003 Glimpses into the Dialectics of Antebellum Landscape Nucleation in Agrarian Michigan. *Journal of Archaeological Method and Theory* 10(4): 369–432.

2014 *A Desolate Place for a Defiant People.* University Press of Florida, Gainesville.

2015 Alienation, Praxis and Significant Social Transformation through Historical Archaeology. In *Historical Archaeologies of Capitalism,*

2nd ed., edited by Mark Leone and Jocelyn Knauf. Springer, New York.

Schivelbusch, Wolfgang
1977 *The Railway Journey: The Industrialization of Time and Space in the 19th Century.* University of California Press, Berkeley.

Schroeder, Jeanne L., and David Gray Carlson
2000 Review: Kenneth Starr: Diabolically Evil? *California Law Review* 88(2): 653–683.

Schwartz, Hillel
1996 *The Culture of the Copy.* Zone Books, New York.

Scott, Doug
2011 Analysis of Ammunition Components from the Lattimer Massacre Site. Report to the Lattimer Massacre Project, College Park, Maryland; Lincoln, Nebraska.

Scott, James C.
2013 *Two Cheers for Anarchism: Six Easy Pieces on Autonomy, Dignity, and Meaningful Work and Play.* Princeton University Press, Princeton, New Jersey.

Self, Robert O.
2003 *American Babylon.* Princeton University Press, Princeton, New Jersey.

Shackel, Paul A.
1993 *Personal Discipline and Material Culture.* University of Tennessee Press, Knoxville.
1996 *Culture Change and the New Technology: An Archeology of the Early American Industrial Era.* Plenum, New York.
2004 Labor's Heritage: Remembering the American Industrial Landscape. *Historical Archaeology* 38: 44–58.
2009 *The Archaeology of American Labor and Working Class Life.* University Press of Florida, Gainesville.
2013 Changing the Past for the Present and the Future. *Historical Archaeology* 47(3): 1–11.

Shackel, Paul, and Michael Roller
2012 The Gilded Age Wasn't So Gilded in the Anthracite Region of Pennsylvania. *International Journal of Historical Archaeology* 16(4): 761–775.

Shackel, Paul, Michael Roller, and Kristin Sullivan
2011 Historical Amnesia and the Lattimer Massacre. *Society for Historical Anthropology Newsletter.* Electronic Document, http://sfaanews.sfaa.net/2011/05/01/historical-amnesia-and-the-lattimer-massacre/, accessed August 15, 2012.

Sheldon, Roy, and Egmont Arens
1932 *Consumer Engineering: A New Technique for Prosperity.* Harper & Brothers, New York.

Sivilich, Dan

2011 Electronic Phase I Archaeological Survey Conducted at: Lattimer 1897 Massacre Site, Lattimer Road, Lattimer, Luzerne County, Pennsylvania. Report prepared for the Lattimer Massacre Project, College Park, Maryland, by Battlefield Restoration and Archaeological Volunteer Organization (BRAVO), Monmouth, New Jersey.

Smith, Laurajane, Paul Shackel, and Gary Campbell (editors)

2011 *Heritage, Labour and the Working Class.* Routledge, New York.

Society for Historical Archaeology (SHA)

2014 *Machine Made Dating.* Webpage. Electronic document, http://www.sha.org/bottle/machinemadedating.htm, accessed January 23, 2014.

Stolarik, M. Mark

2002 A Slovak Perspective on the Lattimer Massacre. *Pennsylvania History* 96(1):31–41.

Stoler, Ann

2009 *Along the Archival Grain.* Princeton: Princeton University Press, Princeton, New Jersey.

Susman, Warren L.

1973 *Culture and Commitment, 1929–1945.* George Braziller, New York.

Tarde, Gabriel

1903 *The Laws of Imitation.* Henry Holt, New York.

1969 *On Communication and Social Influence: Selected Papers.* University of Chicago Press, Chicago, Illinois.

Tarone, Louis A.

2004 *We Were Here Once.* Citizen, Hazleton, Pennsylvania.

Theodori, Ann E., and Gene L. Theodori

2014 Perceptions of Community and Place and the Migration Intentions of At-Risk Youth in Rural Areas. *Journal of Rural Social Sciences* 29(1): 103–121.

Toscano, Alberto

2017 Notes on Late Fascism. Historical Materialism. Website. http://www.historicalmaterialism.org/blog/notes-late-fascism#%20ftn9, accessed June 1, 2017.

Trachtenberg, Alan

1982 *The Incorporation of America: Culture and Society in the Gilded Age.* Hill and Wang, New York.

Tunbridge, J. E., and Gregory J. Ashworth

1996 *Dissonant Heritage: The Management of the Past as a Resource in Conflict.* Wiley, New York.

Turner, George

1977 The Lattimer Massacre and Its Sources. *Slovakia* 27(50): 9–43.

1984 Ethnic Responses to the Lattimer Massacre. In *Hard Coal, Hard Times: Ethnicity and Labor in the Anthracite Region*, edited by David

L. Salay, pp. 126–152. Anthracite Museum Press, Scranton, Pennsylvania.

2002 The Lattimer Massacre: A Perspective from the Ethnic Community. *Pennsylvania History* 96(1): 11–30.

Twain, Mark M., and Charles Dudley Warnier

1873 *The Gilded Age: A Tale of To-Day.* Bobbs–Merrill, Indianapolis, Indiana.

United States Congress

1911 Immigration of Aliens into the United States, 61st U.S. Congress, 3rd Session, Report No. 2. Government Printing Office, Washington, D.C.

1921 Hearings before the Committee on Manufactures, United States Senate, Sixty-sixth Congress, third session, on S. 4828. Government Printing Office, Washington, D.C.

United States House of Representatives (U.S. HoR)

1889 Labor Troubles in the Anthracite Region of Pennsylvania, 1887–1888. H.R. Report No. 4147. Government Printing Office, Washington, D.C.

1901 Papers Relating to the Foreign Relations of the United States, with the Annual Message of the President Transmitted to Congress December 5, 1898. Government Printing Office, Washington, D.C.

United States Presidents Advisory Committee on Government Housing Policies and Programs

1953 *Recommendations on Government Housing Policies and Programs: A Report of the President's Advisory Committee on Government Housing Policies and Programs.* Government Printing Office, Washington, D.C.

Varsanyi, Monica W.

2011 Neoliberalism and Nativism: Local Anti-Immigrant Policy Activism and an Emerging Politics of Scale. *International Journal of Urban and Regional Research* 35(2): 295–311.

Walker, Francis

1924 The Development of the Anthracite Combination. *Annals of the American Academy of Political and Social Science* 111(1): 234–248.

Walker, Mark

2008 "The Floating Army": Transient Labor in Early Twentieth Century California. Paper presented at the Society for Historical Archaeology Conference, Albuquerque, New Mexico.

Wall, Diane

1994 *The Archaeology of Gender.* Plenum Press, New York.

Wallace, Anthony F. C.

1978 *Rockdale: The Growth of an American Village in the Early Industrial Revolution.* Knopf, New York.

1988 *St. Clair: A Nineteenth-century Coal Town's Experience with a Disaster-prone Industry.* Cornell University Press, Ithaca, New York.

Warne, Frank Julian

1904a *The Slav Invasion and the Mine Workers.* J. P. Lippincott, Philadelphia, Pennsylvania.

1904b Some Industrial Effects of Slav Immigration. *Charities* XIII(10): 223–226.

Westmont, Camille

2014 From Homespun to Machine Made: The Rise of Women Wage-Earners in the Pennsylvania Anthracite Region. Paper presented at the 47th Annual Meeting of the Society for Historical Archaeology, Quebec.

Willey, Malcolm M., and Stuart A. Rice

1933a Communication Agencies and Social Life, United States. In *President's Research Committee on Social Trends.* McGraw-Hill, New York.

1933b *The Agencies of Communication.* McGraw-Hill, New York.

Williams, Raymond

1993 Advertising, the Magic System. In *The Cultural Studies Reader,* 2nd ed., edited by Simon During, pp. 410–423. Routledge, New York.

Williams, Phyllis H.

1969 [1938] *South Italian Folkways in Europe and America: A Handbook for Social Workers, Visiting Nurses, School Teachers, and Physicians.* Yale University Press, New Haven, Connecticut.

Wolensky, Kenneth C.

2003 An Industry on Wheels. In *Sweatshop USA: The American Sweatshop in Historical and Global Perspective*, edited by Daniel Bender and Richard Greenwald, pp. 91–116. Routledge, New York.

Wolensky, Robert. P., and William Hastie, Sr.

2013 *Anthracite Labor Wars.* Canal History and Technology Press, Easton, Pennsylvania.

Wood, Margaret C.

2002 Fighting for our Homes": An Archaeology of Women's Domestic Labor and Social Change in a Working Class, Coal Mining Community, 1900–1930. Unpublished PhD dissertation, Department of Anthropology, Syracuse University, Syracuse, New York.

Wurst, LuAnn

1999 Internalizing Class in Historical Archaeology. *Historical Archaeology* 33(1): 7–21.

Wurst, LuAnn, and Randy McGuire

1999 Immaculate Consumption: A Critique of the "Shop till You Drop" School of Human Behavior. *International Journal of Historical Archaeology*, 3(3): 191–200.

Wurst, LuAnn, and Steven Mrozowski
2014 Toward an Archaeology of the Future. *International Journal of Historical Archaeology*, 18(2): 210–223.
Zeidel, Robert F.
2004 *Immigrants, Progressives, and Exclusion Politics: The Dillingham Commission, 1900–1927.* Northern Illinois University Press, DeKalb.
Zinn, Howard
2003 *A People's History of the United States.* Harper Perennial, New York.
Žižek, Slavoj
1989 *The Sublime Object of Ideology.* Verso, London.
1992 Eastern Europe's Republics of Gilead. In *Dimensions of Radical Democracy,* edited by Chantel Mouffe, pp. 193–207. Verso, London
1997 *The Plague of Fantasies.* Verso, London.
1999 *The Ticklish Subject: The Absent Centre of Political Ontology.* Verso, London.
2001 *Enjoy Your Symptom!: Jacques Lacan in Hollywood and Out.* Routledge, New York.
2008 *Violence.* Picador, New York.
2013 *Less Than Nothing: Hegel and the Shadow of Dialectical Materialism.* Verso, London.

INDEX

Page numbers followed by the letter *i* refer to illustrations and those followed by the letter *t* refer to tables.

Advertisement, 115, 125–127, 127*i*, 129, 196, 201; radio and, 131
Agamben, Giorgio, 18, 19, 28–29, 52, 79, 99–100, 197, 198, 200. *See also* Camp (as ideological space); *Homo Sacer*
Alienation, 28, 116; consumerism and, 114, 124, 134, 190
Ammunition: archaeological analysis of, 70–72; recovered during survey of the Lattimer Massacre, 70–73, 72*i*; used during the Lattimer Massacre (historical accounts), 58, 72
Anarchists, 66, 78, 120, 141; antianarchist sentiments, 69, 74
Anthracite (coal): geology and geography of, 13–14, 33, 34, 37; industrial revolution and, 15, 32–34; modernity, 32–33; properties of, 33
Anthracite industry: child labor and, 39–40; history of early formation, 33–36, 138–139; immigrant labor and, 43–46, 44*t*, 49–50, 52–53, 191–192; landscapes of, 79–80; ownership and operators, 35, 42, 84, 161, 193; processes of coal extraction, 36–40, 101–102 (*see also* Mechanization, coal extraction); transportation monopolies and, 35–36, 52; unionization efforts and, 47–49, 53–54, 57, 154; working conditions, 38
Anthracite region: heritage and memory in, 2–3, 180–183; and the Civil War, 42; immigration to, 14, 43–47; geography of, 34; and Federal assistance, 150, 185
Anti-immigration (xenophobia), 9, 17–18; historic, 44–47, 49, 53–55; in the present, 1, 21–24, 194–197
Architecture, of the anthracite coal region, 79–80, 82–92, 93–94, 113; symbolically racialized, 90–92, 192–193. *See also* Company towns; Suburbanization
Arendt, Hannah, 19, 203
Austro-Hungarian Empire, 43, 66, 67, 91
Autonomists, 115

Banditry, 156–157
Benjamin, Walter: on "The Destructive Character," 29, 163–164, 170, 187; on fashion and mass consumerism, 108, 124, 132; on history, 1, 120, 189–190, 205; on sovereignty and violence, 18
Bertolluci, Bernardo (filmmaker), 7, 136
Biopolitics, 8, 28, 100. *See also* Governmentality; *Homo Sacer*
Bootlegging (coal), 154–156
Borders (of nations), and formation of subjects, 18–19, 20, 26, 79, 201
Brown, Wendy, 27–28, 199
Butler, Judith, 74

Camp (as ideological space), 97–100, 197. *See also* Shanty towns (enclaves)
Campanilismo, 157
CAN DO Corporation, 21, 151, 159, 177, 183–186
Capitalism, 9, 162, 203; archaeology of, 121, 133; commodity cycle and, 140; flexible accumulation and neoliberal forms, 9, 21–22, 24, 185, 199; labor (migrant and otherwise) and, 15, 20, 98, 103, 194–197; logistics and infrastructure and, 122–124; machinic conceptions of, 115–116; mass distribution, 124–125; mechanization and, 32–33, 101–102, 117, 120–122; modernity and, 32; obsoletism and waste, 129–130; physical space and, 79, 97–98, 103, 122, 141–142, 187, 192; product design and fashion cycles, 127–129; subjectivation under, 199; as transnational, 31, 40, 122. *See also* Advertisement; Consumerism; Mechanization
Central American immigrants, 20, 22–24, 197
Chain stores, 124–125, 176–177
Chavez, Cesar, 136, 151, 205–206
Class, 8, 10, 15, 19, 31, 142
Cohen, Lizbeth, 111, 124–125, 157, 159, 160
Commodity cycle, 114–115, 122, 128, 140, 161
Company towns, 11, 16–17, 35, 97–99, 192–193
Consumerism, 105–135, 177, 191, 194–195, 208n1
Contradiction (in history or society), 20, 98–99, 134–135, 143, 205
Critical archaeology, 9, 29–30, 137, 203–204

Deindustrialization, 21, 49, 138–140
Deleuze, Gilles, and Felix Guattari, 9, 114–115, 203
Desire: and the subject, 25–26, 199–201; consumerism and, 109, 111–112, 117, 119, 126–127, 129, 134–135, 201
Dillingham Commission, 43, 45–47, 74, 110, 120

Eastern European immigrants: and enjoyment, desire, and consumption, 109–113, 194; immigration to the United States, 14–15, 41, 43–44, 191; as labor, 14, 52; racial stereotypes of, 69, 92, 109–112; in strikes, 54–56, 66; unionization and, 54, 207n2
Economic decline, 21, 49, 138–140
Enjoyment, 109, 140–141, 160–161, 188, 195
Ethnicity, 15, 44–45, 201

Fascism, 136, 190, 203. *See also* Totalitarianism
Flexible accumulation (post-Fordism), 9, 21–22, 24, 123, 151, 185–186; labor markets and, 21, 150, 188, 196. *See also* Neoliberalism
Foodways, communal solidarity and, 152–154
Forgetting (as ideological process), 20, 108, 190, 204. *See also* Memory
Freedom: negative freedom within governmentality, 164, 176, 180, 190, 202–203; in neoliberalism, 25–26, 28, 197; "double," of the proletariat, 19–20, 27–28

Gardens, 78, 87, 94, 152–154, 169, 172, 180
Gilded Age, 32–33
Governmentality, 8, 9, 25, 27–29, 175, 201–203
Gramsci, Antonio, 137
Great Depression, 132, 150, 154–155. *See also* New Deal

Haraway, Donna, 20–21
Hazleton (city), 12*i*; economy, 123, 150; history and settlement, 35, 36, 82*i*; immigration politics today, 21–24, 197; urban renewal and, 163, 170–171, 171*i*, 176–177, 178. *See also* CAN DO Corporation; Illegal Immigration Relief Act (IIRA)
Heritage: labor, 204; dark or difficult, 1; as justificatory evocation for discrimination, 23–24

Heterotopias, 97–100
Homo Sacer, 18, 100

Identity: community, 142, 145, 157–160 (*see also* Camp); in conflict with governmentality, 175; political or national, 15, 18, 20, 52, 74, 79
Ideology: archaeological study of, 16; consumerism and, 108, 114, 126, 130; historical narratives and, 20, 198, 204; of identity and subjects, 18, 27, 52, 79, 201; neoliberalism and, 27, 197, 199; urban renewal and, 164; Žižek and, 200
Illegal Immigration Relief Act (IIRA), 22–23, 197. *See also* Hazleton, immigration politics today
Immigration: and contemporary Hazleton politics, 22–24; labor and, 14, 17–18, 40–46, 53–54, 101–102; the New Immigration, 14–15, 43–44; politics of sovereign nations and, 17–20, 26, 79; spatiality in the company towns and, 84, 86–92; from Western Europe, 57–58. *See also* Anti-immigration (xenophobia); Dillingham Commission; Eastern European immigrants; Italian immigrants
Imperialism, 66–67, 74–75, 75*i*
Industrial Revolution. *See* Anthracite (coal), industrial revolution; Modernity
Infrastructure, 147–148, 149*i*
Interwar Period, 116–120, 122, 125, 126
Irish (immigration), 24, 42–43, 84; Molly Maguires, 43
Irony, in history, 17, 20–21, 105
Italian immigrants, 24, 43–44, 44*i*, 49, 53; participation during the Lattimer strike, massacre, and aftermath, 55–57, 59, 62, 64–65; racial stereotypes of, 91, 110–111; residence and community in Lattimer and Pardeesville, 84–87, 137, 152–154, 156–160; as strikebreakers, 49; in stripping labor, 101

Kafka, Franz, 25, 199–201
Kristeva, Julia, 18, 52, 79, 141

Labor history, 30, 204–205
Lacan, Jacques, 140–141, 160, 199–200
Landscape, 11, 13, 16, 35; capitalist, 16, 142; company town, 77–104, 192–193; mining, 79–80, 181, 184–185; postwar changes to Pardeesville, 164–169; shanty enclave, 13; urban renewal and, 174–176, 180, 187–188, 196
Lattimer Massacre, 51–69; archaeology of, 69–73, 72*i*, 73*i*; contemporary responses to, 63–67; events leading up to, 52–59; scholarship of, 207n1; trial, 67–69; and unionization efforts, 56–57, 59, 138
Lattimer Mines: formation and early history, 47–49, 81–82, 101–102, 106, 107*i*, 113, 139–160, 164–169; history post–World War II, 164–169; post 1960s, 177–180; landscape and architecture of, 82–92, 83*i*, 88*i*, 89*i*, 145, 146*i*, 152, 164–169
Lazzarato, Maurizio, 103, 115, 116, 135, 192, 202
Leone, Mark, 16, 17, 79, 133, 198
Logistics, 122–124
Longue durée, 80, 161, 193; perspective of the twentieth century, 6, 103, 120, 191, 194

Machine Age. *See* Interwar Period
Machinic (theory), 109, 114–116, 142, 175; and consumerism, 120–134, 194–195
Manufacturing processes. *See* Mechanization
Marx, Karl, 15, 19, 27, 41, 114–116, 189
Mass consumerism. *See* Consumerism
Mass production, 33, 108, 114, 116–117, 121–122, 133, 190. *See also* Consumerism; Mechanization
Matthews, Christopher, 6, 11, 27, 103, 116, 198, 199
Mechanization, 120–121; coal extraction and, 36, 52–53, 100–102, 138; deskilling of labor and, 32–33, 41, 126; glass production, 121; and middle class, formation of, 193; in Pardee Mines, 82, 102; and surplus production, 108, 117, 195
Media (electronic communicative), political implications, 109, 114, 130–133, 203–204

Memory, 2–3, 29–30, 136, 204–205; commodification and fashion, 108, 134, 190; urban renewal and, 187. *See also* Forgetting
Middle class, 35–36, 100, 117–118, 128, 195–196
Migrants. *See* Immigration
Modernity: and coal, 32–33, 124; and company towns, 98, 100; late forms, 18–19, 202; mass consumerism and, 108–109, 128

Nationalism, 9. *See also* Borders (of nations); Patriotism
Neoliberalism: Hazleton's economy and, 21, 24, 196–197; as subjectivating ideology, 24–28, 201–202. *See also* Flexible accumulation
New Deal, and the anthracite region, 40, 150
New Immigration, 14–15, 43–44. *See also* Eastern European immigrants; Immigration; Italian immigrants

Obsoletism, 129–130

Pardee family, 35, 47–49, 53, 59, 81, 101–102, 139, 145
Pardeesville (i.e., Lattimer No. 2), early history of, 17, 81–84, 82*i*, 87–92, 88*i*, 89*i*, 113, 141–145. *See also* Lattimer Mines
Patch towns, 11, 76, 80, 152, 163–164, 175, 192. *See also* Company towns
Patriotism, 42–43; the American flag as symbol of, 60, 68–69, 105–107, 107*i*
Permissive technologies, 123–124, 186–187
Plastic, 113, 127–129
Post-Fordism. *See* Flexible accumulation; Neoliberalism
Precarity, 19, 24, 52, 79, 186, 192
Proletarianization, 14–15, 19, 191
Psychoanalysis, 117, 125–126, 140

Racism, 11, 46, 124
Radio, political employment of, 130–133, 203

Shackel, Paul, 29–30, 69, 79, 133, 204
Shanty towns (enclaves), 86–97, 99–100, 102–103, 113, 194; urban renewal, as subject of, 159, 164–169, 196–198; *See also* Camp (as ideological space); Company towns; Urban renewal
Shopping centers, 124–125, 176–177
Slavs. *See* Eastern European immigrants
Sovereignty, 18–19, 28, 74, 79, 100. *See also* Borders (of nations)
Strikes: precipitating the Lattimer Massacre, 10, 49, 55–59, 68, 205; following the Massacre, 64–66, 138; elite's response to, 42, 47–48; mechanization and, 53, 101; and gardens, 153
Subjectivation, 9, 24–29, 114, 116, 133–135, 164, 175, 198–203
Suburbanization, 12–13, 16*i*, 77, 174–176, 181; architecture and aesthetics of, 84, 92–93, 113, 169
Surplus: as the subject of archaeology, 3, 9, 137, 161, 201; desire, 111, 194, 201; enjoyment, 140–141, 160–161, 188, 195; labor, 14–15, 40, 49, 52, 101, 103, 123, 186, 192; life, 78; parts, 180; production, 5, 98, 101, 108, 190, 195; value, 76, 114–115, 140

Textile industry, 123, 150, 184–185
Totalitarianism, 190. *See also* Fascism
Tourism, 181–183
Transnational history, 13, 17, 31, 40, 66
Twentieth century, history of, 1–3, 17–20, 25–29, 100, 198–204

UMWA (United Mine Workers of America): early efforts at unionization in the anthracite region, 47, 49, 53–54, 56; last quarter of the twentieth century, 188, 196; role in the Lattimer Massacre, 56–57, 59; role in the memorialization of the Lattimer Massacre, 151; role in the period of employee ownership at Lattimer Mines, 143–144
Unions. *See* UMWA
Urban redevelopment. *See* Urban renewal
Urban renewal, 163, 170–177, 196, 202

Violence, 8, 10–11, 74–75, 77; structural or systemic, 2, 79, 191–194; symbolic, 15–16, 23, 28, 54; subjective violence, sovereignty and, 18, 20
Vogelfrei, 19, 41. *See also* Proletarianization; Surplus, labor

Women, and labor politics, 56, 65, 76; and work, 123, 150, 154

Xenophobia. *See* Anti-immigration

Žižek, Slavoj: on *Blue Velvet*, 77; on desire, enjoyment, and political power, 28, 109, 141, 200; on "the Real," 160–161; on capitalism, 103, 162; on violence, 10, 77, 164

MICHAEL P. ROLLER is a research affiliate of the Department of Anthropology at the University of Maryland. His research interests are in the intersection of historical archaeology, critical theory, labor history, and migration. He is the author of "Diabolical Consumerism: Mass Psychology and Social Production between the Gilded and the Golden Ages," in *Historical Archaeologies of Capitalism*, edited by Mark Leone and Jocelyn Knauf, and "Rewriting Narratives of Labor Violence: A Transnational Perspective of the Lattimer Massacre," in *Historical Archaeology*. He is currently employed as an archaeologist for the National Park Service.

CULTURAL HERITAGE STUDIES

Edited by Paul A. Shackel, University of Maryland

Heritage of Value, Archaeology of Renown: Reshaping Archaeological Assessment and Significance, edited by Clay Mathers, Timothy Darvill, and Barbara J. Little (2005)

Archaeology, Cultural Heritage, and the Antiquities Trade, edited by Neil Brodie, Morag M. Kersel, Christina Luke, and Kathryn Walker Tubb (2006)

Archaeological Site Museums in Latin America, edited by Helaine Silverman (2006)

Crossroads and Cosmologies: Diasporas and Ethnogenesis in the New World, by Christopher C. Fennell (2007)

Ethnographies and Archaeologies: Iterations of the Past, by Lena Mortensen and Julie Hollowell (2009)

Cultural Heritage Management: A Global Perspective, by Phyllis Mauch Messenger and George S. Smith (2010; first paperback edition, 2014)

God's Fields: An Archaeology of Religion and Race in Moravian Wachovia, by Leland Ferguson (2011; first paperback edition, 2013)

Ancestors of Worthy Life: Plantation Slavery and Black Heritage at Mount Clare, by Teresa S. Moyer (2015)

Slavery Behind the Wall: An Archaeology of a Cuban Coffee Plantation, by Theresa A. Singleton (2015; first paperback edition, 2016)

Excavating Memory: Sites of Remembering and Forgetting, edited by Maria Theresia Starzmann and John R. Roby (2016)

Mythic Frontiers: Remembering, Forgetting, and Profiting with Cultural Heritage Tourism, by Daniel R. Maher (2016)

Critical Theory and the Anthropology of Heritage, by Melissa F. Baird (2018)

Heritage at the Interface: Interpretation and Identity, edited by Glenn Hooper (2018)

Cuban Cultural Heritage: A Rebel Past for a Revolutionary Nation, by Pablo Alonso González (2018)

The Rosewood Massacre: An Archaeology and History of Intersectional Violence, by Edward González-Tennant (2018)

Race, Place, and Memory: Deep Currents in Wilmington, North Carolina, by Margaret M. Mulrooney (2018)

An Archaeology of Structural Violence: Life in a Twentieth-Century Coal Town, by Michael P. Roller (2018)

www.ingramcontent.com/pod-product-compliance
Lightning Source LLC
LaVergne TN
LVHW091529200926
848483LV00002B/337